Collins

T0173632

Edexcel A-Level

Maths
Year 1 & AS

Revision Guide

Rebecca Evans, Leisa Bovey
and Phil Duxbury

About this Revision & Practice book

Revise

These pages provide a recap of everything you need to know for each topic.

You should read through all the information before taking the Quick Test at the end. This will test whether you can recall the key facts.

Quick Test

1. Expand and simplify $3x(x^2 - 2y + 6) - 2y(x + 2) + 3xy$.
2. Expand and simplify $(2x + 3)(x - 2)(x + 1)$.
3. Factorise completely $3x^2 - 10x - 8$.
4. Factorise completely $6x^2 - 23x + 20$.
5. Factorise completely $2x^3 + 8x^2 + 6x$.

Practise

These topic-based questions appear shortly after the revision pages for each topic and will test whether you have understood the topic. If you get any of the questions wrong, make sure you read the correct answer carefully.

Review

These topic-based questions appear later in the book, allowing you to revisit the topic and test how well you have remembered the information. If you get any of the questions wrong, make sure you read the correct answer carefully.

Mix it Up

These pages feature a mix of questions for all the different topics, just like you would get in an exam. They will make sure you can recall the relevant information to answer a question without being told which topic it relates to.

Test Yourself on the Go

Visit our website at **collins.co.uk/collinsalevelrevision** and print off a set of flashcards. These pocket-sized cards feature questions and answers so that you can test yourself on all the key facts anytime and anywhere. You will also find lots more information about the advantages of spaced practice and how to plan for it.

Workbook

This section features even more topic-based questions as well as practice exam papers, providing two further practice opportunities for each topic to guarantee the best results.

ebook

To access the ebook revision guide visit

collins.co.uk/ebooks

and follow the step-by-step instructions.

Contents

Statistics and Mechanics

Indices and Surds

You must be able to:

- Use the laws of indices
- Manipulate surds
- Rationalise the denominator.

Indices

Base $\longrightarrow a^m \longleftarrow$ Index

- **Index** (plural **indices**) is another word for power or exponent, and is the number of times a base number is multiplied by itself.
- The **base** is the number that is being multiplied by itself.
- For example, 3^4 means to multiply 3 (the base) by itself 4 (the index) times:

  $3^4 = 3 \times 3 \times 3 \times 3$
- The laws of indices are shown in this table:

$a^m \times a^n = a^{m+n}$	$(ab)^n = a^n b^n$	$a^{-m} = \dfrac{1}{a^m}$
$a^m \div a^n = a^{m-n}$	$a^{\frac{1}{n}} = \sqrt[n]{a}$	$a^1 = a$
$(a^m)^n = a^{mn}$	$a^{\frac{m}{n}} = \sqrt[n]{a^m}$	$a^0 = 1$

> **Key Point**
>
> The rules of indices only apply with the same base.

> **Key Point**
>
> You must be able to use all the laws of indices together.

1) Simplify $3k^2 \times 5k^6$.

$= 3 \times 5 \times k^2 \times k^6$
$= 15 \times k^{2+6}$
$= 15k^8$

2) Simplify $(2m^7)^3 \div m^2$.

$= 2^3 \times (m^7)^3 \div m^2$
$= 8 \times m^{21} \div m^2$
$= 8 \times m^{21-2}$
$= 8m^{19}$

3) Simplify $\left(4x^8 y^{-6}\right)^{\frac{3}{2}}$.

$= 4^{\frac{3}{2}} \times x^{8 \times \left(\frac{3}{2}\right)} \times y^{-6 \times \left(\frac{3}{2}\right)}$
$= \sqrt{4^3} \times x^{12} \times y^{-9}$
$= \dfrac{8x^{12}}{y^9}$

Surds

- A **surd** is a square **root** that cannot be further simplified, e.g. $\sqrt{3}, 3\sqrt{10}$.
- Surds are **irrational numbers** (i.e. they are not **rational numbers**) so they cannot be written as a fraction $\frac{a}{b}$ where a and b are integers.
- To manipulate surds, remember:

$\sqrt{xy} = \sqrt{x} \times \sqrt{y}$	$\sqrt{\dfrac{x}{y}} = \dfrac{\sqrt{x}}{\sqrt{y}}$	$\left(\sqrt{x}\right)^2 = x$

> **Key Point**
>
> $\sqrt{n}$ of any positive number, n, that is not a square number, is a surd.

1) Simplify $\sqrt{50}$.

$= \sqrt{25} \times \sqrt{2} = 5\sqrt{2}$

2) Expand $(2 + \sqrt{7})(3 - \sqrt{2})$.

$= 6 - 2\sqrt{2} + 3\sqrt{7} - \sqrt{14}$

3) Simplify $\sqrt{48} + \sqrt{147} - \sqrt{108}$.

$= \sqrt{16 \times 3} + \sqrt{49 \times 3} - \sqrt{36 \times 3}$

$= 4\sqrt{3} + 7\sqrt{3} - 6\sqrt{3}$

$= 5\sqrt{3}$

> **Key Point**
>
> Remember
> $(a + b)^2 \neq a^2 + b^2$
> $\sqrt{x + y} \neq \sqrt{x} + \sqrt{y}$

Rationalising the Denominator

- When a surd is in the denominator of a fraction, you must rationalise the denominator, i.e. use equivalent fractions to rewrite the fraction without a surd in the denominator.

To rationalise the denominator of:	Multiply by:
$\dfrac{1}{\sqrt{a}}$	$\dfrac{\sqrt{a}}{\sqrt{a}}$
$\dfrac{1}{a + \sqrt{b}}$	$\dfrac{a - \sqrt{b}}{a - \sqrt{b}}$
$\dfrac{1}{a - \sqrt{b}}$	$\dfrac{a + \sqrt{b}}{a + \sqrt{b}}$

> **Key Point**
>
> Multiplying a number by a fraction that has the same expression in the numerator and the denominator is the same as multiplying by 1.

Rationalise the denominator: $\dfrac{2}{3 + \sqrt{3}}$

$= \dfrac{2}{3 + \sqrt{3}} \times \dfrac{3 - \sqrt{3}}{3 - \sqrt{3}} = \dfrac{2(3 - \sqrt{3})}{(3 + \sqrt{3})(3 - \sqrt{3})} = \dfrac{6 - 2\sqrt{3}}{9 - 3} = \dfrac{6 - 2\sqrt{3}}{6} = \dfrac{3 - \sqrt{3}}{3}$

> **Key Point**
>
> $(a + \sqrt{b})(a - \sqrt{b}) = a^2 - b$

> **Key Words**
>
> index
> base
> surd
> root
> irrational number
> rational number

> **Quick Test**
>
> 1. Simplify $x^6 \div 2x^3$.
> 2. Expand and simplify $(3 + \sqrt{2})(2 + \sqrt{8})$.
> 3. Rationalise the denominator: $\dfrac{3}{3 - \sqrt{5}}$

Manipulating Algebraic Expressions

You must be able to:

- Expand brackets and collect like terms of two or three expressions
- Factorise expressions by pulling out the greatest common factor
- Factorise quadratic expressions $ax^2 + bx + c$
- Factorise cubic expressions $ax^3 + bx^2 + cx$.

Expanding Brackets of Two Expressions

- Multiply each **term** in the first **expression** by each term in the second expression.
- After finding the **product**, collect like terms to **simplify** the resulting expression.

  Expand the brackets and simplify:

  a) $3(x^2 - 4x + 6) - 2x(x - 8) + 5$

  $= 3x^2 - 12x + 18 - 2x^2 + 16x + 5$

  $= 3x^2 - 2x^2 - 12x + 16x + 18 + 5$

  $= x^2 + 4x + 23$

  b) $(2x + 3)(x - 2y + 4)$

  $= 2x^2 - 4xy + 8x + 3x - 6y + 12$

  $= 2x^2 - 4xy + 11x - 6y + 12$

> **Key Point**
>
> Watch out for the signs of each term. Remember that $- \times - = +$ and remember to multiply a negative in front of a bracket through each term in the bracket.

> **Key Point**
>
> Make sure to multiply each term in the first expression by each term in the other expression.

Expanding Brackets of Three Expressions

- Multiply any two brackets together then multiply the product by the remaining bracket.

  Expand the brackets and simplify:

  $(x + 1)(x - 2)(x + 3)$

  $= (x^2 - 2x + x - 2)(x + 3)$

  $= (x^2 - x - 2)(x + 3)$

  $= x^3 + 3x^2 - x^2 - 3x - 2x - 6$

  $= x^3 + 2x^2 - 5x - 6$

> **Key Point**
>
> Remember to multiply each term by each other term.

Factorising using Common Factors

- To factorise means to write an expression as a product.
- Factorising is the reverse of expanding brackets.

  Factorise completely $3x^2 - 6xy$.

  $= 3x(x - 2y)$

> **Key Point**
>
> To factorise, look for the greatest common factor of each term in the expression and pull it outside the brackets.

Factorising Quadratic Expressions $ax^2 + bx + c$

- To factorise a quadratic expression in the form $ax^2 + bx + c$:
  - identify the value of a, b and c
  - find two numbers whose product is $a \times c$ and that add to b
  - rewrite bx as a sum using these two numbers
  - factorise each pair of terms by pulling out the common factors
  - write the expression as a product of two brackets by pulling out the common factor (the expression inside the brackets).

> **Key Point**
>
> When factorising each pair of terms, make sure the expressions inside the brackets are the same.

1) Factorise completely $2x^2 - 5x - 12$.

In this case, $a = 2$, $b = -5$ and $c = -12$
$a \times c = 2 \times -12 = -24$ and $b = -5$
Two numbers that multiply to -24 and add to -5 are 3 and -8.
$2x^2 - 5x - 12 = 2x^2 + 3x - 8x - 12$
$2x^2 + 3x - 8x - 12 = x(2x + 3) - 4(2x + 3)$ ← Factorise each pair of terms.
Each term has a common factor of $(2x + 3)$.
$x(2x + 3) - 4(2x + 3) = (2x + 3)(x - 4)$

2) Factorise completely $2x^2 - 4x - 6$.
$= 2(x^2 - 2x - 3)$
Factorising $x^2 - 2x - 3$
$= x^2 - 3x + x - 3$
$= x(x - 3) + 1(x - 3)$
$= (x - 3)(x + 1)$
Therefore:
$2x^2 - 4x - 6 = 2(x - 3)(x + 1)$

> **Key Point**
>
> Remember to check if there is a common factor of each term first.

3) Factorise completely $16x^4 - 25y^2$.
$16x^4 - 25y^2 = (4x^2 + 5y)(4x^2 - 5y)$ ← Use the difference of two squares.

Factorising Cubic Expressions $ax^3 + bx^2 + cx$

- First pull out the common factor of ax, bx and cx.
- Factorise the resulting quadratic expression $ax^2 + bx + c$.

Factorise completely $3x^3 + 3x^2 - 6x$.
$= 3x(x^2 + x - 2)$
$= 3x(x + 2)(x - 1)$

> **Key Point**
>
> Remember the difference of two squares:
> $a^2 - b^2 = (a + b)(a - b)$

Quick Test

1. Expand and simplify $3x(x^2 - 2y + 6) - 2y(x + 2) + 3xy$.
2. Expand and simplify $(2x + 3)(x - 2)(x + 1)$.
3. Factorise completely $3x^2 - 10x - 8$.
4. Factorise completely $6x^2 - 23x + 20$.
5. Factorise completely $2x^3 + 8x^2 + 6x$.

> **Key Words**
>
> term
> expression
> product
> simplify

Expanding and Dividing Polynomials

You must be able to:

- Use Pascal's Triangle and the Binomial Theorem to expand a binomial $(a + bx)^n$
- Divide polynomials and simplify algebraic fractions
- Use the Factor Theorem to identify factors of f(x) and to factorise cubic expressions.

Expanding Binomials

- Use **Pascal's Triangle** or the Binomial Theorem (**binomial expansion**) to find the coefficients of the expansion of a **binomial** $(a + bx)^n$.
- Pascal's Triangle is the pattern formed by the coefficients of these expansions:

$(a+b)^0 =$ 1   1

$(a+b)^1 = a+b$ 1   1

$(a+b)^2 = a^2 + 2ab + b^2$ 1   2   1

$(a+b)^3 = a^3 + 3a^2b + 3ab^2 + b^3$ 1   3   3   1

$(a+b)^4 = a^4 + 4a^3b + 6a^2b^2 + 4ab^3 + b^4$ 1   4   6   4   1

$(a+b)^5 = a^5 + 5a^4b + 10a^3b^2 + 10a^2b^3 + 5ab^4 + b^5$ 1   5   10   10   5   1

- Binomial expansion:

$$(a + b)^n = a^n + \binom{n}{1}a^{n-1}b + \binom{n}{2}a^{n-2}b^2 + \binom{n}{3}a^{n-3}b^3 + ... + \binom{n}{n-1}ab^{n-1} + b^n$$

> **Key Point**
>
> You can use your calculator to find the coefficients of the binomial expansion by calculating **nCr** where **n** is the power and **r** is the term number.

1) Write down the expansion of $(2 + x)^4$.

$(2 + x)^4 = 2^4 + (4 \times 2^3 \times x) + (6 \times 2^2 \times x^2) + (4 \times 2 \times x^3) + x^4$

$= 16 + 32x + 24x^2 + 8x^3 + x^4$

Look at the fifth row of Pascal's Triangle to find the coefficients of the expansion.

2) Determine the coefficient of x^4 in the expansion of $(2x - 1)^{14}$.

Using the Binomial Theorem, the coefficient is $\binom{14}{10}2^4(-1)^{10}$

$= 1001 \times 16 \times 1 = 16016$

*Use your calculator to find $\binom{14}{10}$ using the nCr button where **n** = 14 and **r** = 10.*

3) a) Using the first four terms of the binomial expansion of $\left(1 - \frac{x}{5}\right)^{15}$, find an approximation for the value of 0.96^{15}.

$\binom{15}{0}1^{15}\left(-\frac{x}{5}\right)^0 + \binom{15}{1}1^{14}\left(-\frac{x}{5}\right)^1 + \binom{15}{2}1^{13}\left(-\frac{x}{5}\right)^2 + \binom{15}{3}1^{12}\left(-\frac{x}{5}\right)^3$

The first four terms.

$= 1 - 3x + \frac{21}{5}x^2 - \frac{91}{25}x^3$

$1 - \frac{x}{5} = 0.96, -\frac{x}{5} = -0.04$, so $x = 0.2$

To find an approximation for the value of 0.96, solve $1 - \frac{x}{5} = 0.96$ for x.

$1 - 3x + \frac{21}{5}x^2 - \frac{91}{25}x^3 = 1 - (3 \times 0.2) + \left(\frac{21}{5} \times 0.2^2\right) - \left(\frac{91}{25} \times 0.2^3\right)$

Substituting $x = 0.2$.

$= 0.53888$

b) Find the percentage error in the estimation.

% error $= \dfrac{\text{error}}{\text{actual value}} \times 100 = \dfrac{0.96^{15} - 0.53888}{0.96^{15}} \times 100 = 0.59\%$

Algebraic Fractions

- **Algebraic fractions** are fractions with algebraic expressions in the numerator and/or the denominator.
- To simplify algebraic fractions, factorise the numerator and denominator where possible and cancel common factors.

Simplify $\dfrac{x^2 + 2x - 35}{2x^2 - 7x - 15}$

$$= \frac{(x + 7)(x - 5)}{(2x + 3)(x - 5)} = \frac{x + 7}{2x + 3}$$

Dividing Polynomials

> **Key Point**
>
> Use long division to divide polynomials.

- Use long division to divide when both the **dividend** and the **divisor** are **polynomials**. The **quotient** will also be a polynomial. The remainder may be a constant or a polynomial.

1) Divide $2x^3 + 5x^2 - 11x + 4$ by $2x - 1$.

$$
\begin{array}{r}
x^2 + 3x - 4 \\
2x - 1 \overline{)\ 2x^3 + 5x^2 - 11x + 4} \\
-(2x^3 - x^2) \\
\hline
6x^2 - 11x \\
-(6x^2 - 3x) \\
\hline
-8x + 4 \\
-(-8x + 4) \\
\hline
0
\end{array}
$$

Start by dividing the first term of the dividend by the first term of the divisor.

Multiply the resulting term by the divisor and subtract the product from the dividend.

Bring down the next term of the dividend.

Divide the first term of the resulting polynomial by the first term of the dividend and repeat the previous steps until you cannot divide any further.

Remember to include $0x^3$ when writing the dividend.

2) Divide $2x^4 + 3x^2 - x + 3$ by $x - 1$.

$$
\begin{array}{r}
2x^3 + 2x^2 + 5x + 4 \\
x - 1 \overline{)\ 2x^4 + 0x^3 + 3x^2 - x + 3} \\
-(2x^4 - 2x^3) \\
\hline
2x^3 + 3x^2 \\
-(2x^3 - 2x^2) \\
\hline
5x^2 - x \\
-(5x^2 - 5x) \\
\hline
4x + 3 \\
-(4x - 4) \\
\hline
7
\end{array}
$$

$(2x^4 + 3x^2 - x + 3) \div (x - 1)$
$= 2x^3 + 2x^2 + 5x + 4$ remainder 7

The Factor Theorem

- For polynomial f(x), $(x - a)$ is a factor of f(x) if $f(a) = 0$.
- Conversely, for polynomial f(x), if $f(a) = 0$, then $(x - a)$ is a factor of f(x).
- The remainder of $f(x) \div (x - b)$ is the value of $f(b)$.

1) Show that $(x - 1)$ is not a factor of $f(x) = 2x^4 + 3x^2 - x + 3$.

$f(1) = 2(1^4) + 3(1^2) - (1) + 3 = 7$

$f(1) = 7$ and so the remainder of $(2x^4 + 3x^2 - x + 3) \div (x - 1)$ is also 7.

2) Show that $x + 3$ is a factor of $2x^3 + 3x^2 - 11x - 6$ and hence fully factorise.

$2x^3 + 3x^2 - 11x - 6$

$f(-3) = 2(-3^3) + 3(-3^2) - 11(-3) - 6 = -54 + 27 + 33 - 6 = 0$

Therefore $(x + 3)$ is a factor.

Long division gives $(2x^3 + 3x^2 - 11x - 6) \div (x + 3) = 2x^2 - 3x - 2$

Factorising $2x^2 - 3x - 2 = (2x + 1)(x - 2)$

Therefore, $2x^3 + 3x^2 - 11x - 6 = (x + 3)(2x + 1)(x - 2)$

> **Quick Test**
>
> 1. Write down the first four terms in the expansion of $(1 + 2x)^{12}$ in ascending powers of x.
> 2. Simplify $\dfrac{x^2 + 2x - 3}{2x^2 + 10x + 12}$
> 3. $f(x) = 2x^3 + x^2 - 7x - 6$
>   a) Using the Factor Theorem, show that $(x + 1)$ is a factor of f(x).
>   b) Divide f(x) by $(x + 1)$.
>   c) Fully factorise $2x^3 + x^2 - 7x - 6$.

> **Key Words**
>
> Pascal's Triangle
> binomial expansion
> binomial
> algebraic fraction
> polynomial
> dividend
> divisor
> quotient

Quadratic Equations

You must be able to:

- Solve quadratic equations by factorising, completing the square and using the quadratic formula
- Identify the roots of a quadratic function
- Graph quadratic functions.

Solving Quadratic Equations

Factorising

- To solve by factorising, set each bracket equal to 0 and solve for x.

Solve $3x^2 - 7x + 1 = 3 - 2x$.

$3x^2 - 5x - 2 = 0$ — Simplify and write the equation in the form $ax^2 + bx + c = 0$.

$(3x + 1)(x - 2) = 0$ — Factorise the expression on the left-hand side (see pages 8–9 for factorisation).

$3x + 1 = 0$ or $x - 2 = 0$, so $x = -\frac{1}{3}$ or $x = 2$

Completing the Square

Solve $2x^2 + 4x + 5 = 8 + 2x$.

$2x^2 + 2x - 3 = 0$ — Simplify and write the equation in the form $ax^2 + bx + c = 0$.

$x^2 + x - \frac{3}{2} = 0$ — Divide every term by 'a' if $a \neq 1$.

$x^2 + x = \frac{3}{2}$ — Move the constant term to the right-hand side.

$x^2 + x + \left(\frac{1}{2}\right)^2 = \frac{3}{2} + \left(\frac{1}{2}\right)^2$ — Add $\left(\frac{b}{2}\right)^2$ to both sides.

$\left(x + \frac{1}{2}\right)^2 = \frac{7}{4}$ — Factorise the left-hand side and simplify the right-hand side.

$x + \frac{1}{2} = \pm\frac{\sqrt{7}}{2}$ — Take the square root of both sides and solve for x.

$x = -\frac{1}{2} \pm \frac{\sqrt{7}}{2}$

$x = \frac{-1-\sqrt{7}}{2}$ or $x = \frac{-1+\sqrt{7}}{2}$

> **Key Point**
>
> In general, $x^2 + bx + c$
> $= \left(x + \left(\frac{b}{2}\right)\right)^2 - \left(\frac{b}{2}\right)^2 + c$

> **Key Point**
>
> To solve for x by completing the square when $a \neq 1$, divide through by a first.

Using the Quadratic Formula

- The solutions to a **quadratic** equation of the form $ax^2 + bx + c = 0$ can be found using the quadratic formula $x = \dfrac{-b \pm \sqrt{b^2 - 4ac}}{2a}$

Solve $2x^2 - x - 5 = 0$.

$a = 2$, $b = -1$, $c = -5$ — Identify the values of a, b and c and substitute them into the quadratic formula.

$x = \dfrac{-b \pm \sqrt{b^2 - 4ac}}{2a} = \dfrac{-(-1) \pm \sqrt{(-1)^2 - (4 \times 2 \times (-5))}}{2 \times 2} = \dfrac{1 \pm \sqrt{41}}{4}$

$x = \dfrac{1 + \sqrt{41}}{4}$ and $x = \dfrac{1 - \sqrt{41}}{4}$

> **Key Point**
>
> Remember the quadratic formula,
> $x = \dfrac{-b \pm \sqrt{b^2 - 4ac}}{2a}$

Roots and Graphing Quadratic Functions

- The **roots** of a **function** are the **x-intercepts**. A quadratic function can have no real roots, one (repeated) or two real roots. The value of the **discriminant**, $b^2 - 4ac$, determines the number of real roots.

Determine the number of real roots of the function $y = 3x^2 - 5x + 2$.

Using the discriminant, $b^2 - 4ac = (-5)^2 - (4 \times 3 \times 2) = 25 - 24 = 1$

$1 > 0$, so $y = 3x^2 - 5x + 2$ has two real roots.

No real roots: does not cross $y = 0$
$$b^2 - 4ac < 0$$

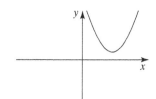

- The curved shape of the graph of a quadratic function is called a **parabola**.
- To graph a quadratic function, identify the y-intercept, the **turning point** and the line of symmetry, and the roots (x-intercepts).

One real root: just touches $y = 0$
$$b^2 - 4ac = 0$$

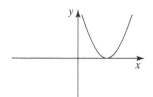

Graph the function $f(x) = 3x^2 - 8x + 4$.

The y-intercept is (0, 4).

$y = 3x^2 - 8x + 4$ ← Find the turning point and the line of symmetry by completing the square (or by using differentiation – see page 58.)

$y = 3\left(x^2 - \left(\frac{8}{3}\right)x + \left(\frac{4}{3}\right)\right)$ ← Factor out the value of a and complete the square of the terms inside the brackets.

$y = 3\left(\left(x - \frac{4}{3}\right)^2 - \frac{4}{9}\right)$

$y = 3\left(x - \frac{4}{3}\right)^2 - \frac{4}{3}$

The turning point is $\left(\frac{4}{3}, -\frac{4}{3}\right)$, the line of symmetry is $x = \frac{4}{3}$

Use any method to find the roots.

$0 = 3x^2 - 8x + 4$

$0 = (3x - 2)(x - 2)$

$x = \frac{2}{3}$ or $x = 2$

The roots are $\left(\frac{2}{3}, 0\right)$ and (2, 0).

Two real roots: crosses $y = 0$ twice
$$b^2 - 4ac > 0$$

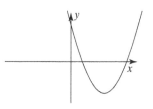

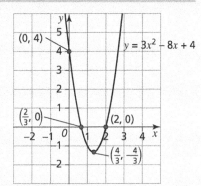

Sketch the graph using symmetry and the critical values found.

Key Point

When the function is in the form $f(x) = p(x + q)^2 + r$, the turning point is at $(-q, r)$.

Using Quadratic Equations in Context

A ball is launched from an initial height of 10 m with an initial velocity of 19.6 ms⁻¹. Find the maximum height and the time at which it reaches that height, given the formula for the height of a ball as a function of time is $h(t) = -4.9t^2 + v_0 t + h_0$, where t is the time in seconds, v_0 is the initial velocity and h_0 is the initial height in metres.

$h(t) = -4.9t^2 + 19.6t + 10$

$h(t) = -4.9(t - 2)^2 + 29.6$ ← Complete the square.

The ball reaches its maximum height of 29.6 m at two seconds. ← The maximum value is at (2, 29.6).

Key Words

quadratic
root (of a function)
function
discriminant
parabola
turning point

Quick Test

1. Solve $x^2 - 5x + 14 = 12 + 3x$ by completing the square.
2. Solve $3x^2 + 2x + 10 = 16$ using the quadratic formula.
3. Draw a graph of the quadratic function $y = 2x^2 - 8x + 13$.

Simultaneous Equations

You must be able to:

- Solve simultaneous equations involving two linear equations in two variables using elimination and substitution
- Solve simultaneous equations involving one linear and one non-linear equation in two variables using substitution
- Represent solutions to simultaneous equations graphically.

Simultaneous Linear Equations

- To solve a set of simultaneous equations means to find a solution set that satisfies both equations.
- Simultaneous linear equations can be solved by the elimination method and by the substitution method.

Solve the simultaneous equations $3x + 2y = 11$ and $5y = x - 15$ using the substitution method and using the elimination method.

Using substitution:

Rearrange $5y = x - 15$ to $x = 5y + 15$

Substitute $x = 5y + 15$ into $3x + 2y = 11$

$3(5y + 15) + 2y = 11$

$15y + 45 + 2y = 11$

$17y = -34$

$y = -2$

Using elimination:

$3x + 2y = 11$ **(1)** ← Label each equation.

$5y = x - 15$ **(2)**

Rearrange equation **(2)**.

$-x + 5y = -15$ **(3)** ← Rearrange the equations so that both variables are on the same side.

Multiply equation **(3)** by 3:

$-3x + 15y = -45$ **(4)** ← Multiply one or both equations by a constant so that the coefficient of one variable is the same in both equations.

Add equations **(1)** and **(4)**.

$3x + 2y = 11$
$-3x + 15y = -45$ ← If the Signs of the variables you are eliminating are the Same, Subtract one equation from the other (remember SSS). If the signs are not the same, add the equations.
$\overline{17y = -34}$
$y = -2$

Substituting $y = -2$ into either of the original equations finds $x = 5$.

The solution is $x = 5$, $y = -2$.

Linear and Non-linear Equations

- Simultaneous equations involving one linear and one non-linear equation can be solved using the substitution method.

Find the solutions to $2y - x = 10$ and $x^2 + xy - 2y^2 = 0$.

$(2y - 10)^2 + y(2y - 10) - 2y^2 = 0$ ← Rearrange the linear equation and substitute into the non-linear equation.

$4y^2 - 50y + 100 = 0$ ← Solve the quadratic by any method.

$2(2y^2 - 25y + 50) = 0$

$(y - 10)(2y - 5) = 0$

$y = 10$, $y = \frac{5}{2}$

$(2 \times 10) - x = 10$, $x = 10$ ← Substitute the y-values into the linear equation and solve to find the corresponding values of x.

$(2 \times \frac{5}{2}) - x = 10$, $x = -5$

The two solutions are $x = 10$, $y = 10$ and $x = -5$, $y = \frac{5}{2}$

> **Key Point**
>
> The solutions to simultaneous equations come in pairs ($x =$ and $y =$). Remember to write the solutions in pairs.

> **Key Point**
>
> Use substitution and then solve the resulting quadratic equation.

Graphs of Simultaneous Equations

- The solutions to simultaneous equations are the point or points where the lines or curves intersect.

Simultaneous linear equations have zero, one or infinite points of intersection

$y = -\frac{2}{3}x + 4$ and $3y = -2x - 4$

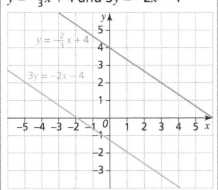

No solutions (parallel lines)

$y = -\frac{3}{4}x + 4$ and $y + 4 = 3x$

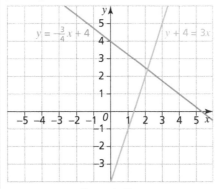

One solution

$2y = 4x - 2$ and $y = 2x - 1$

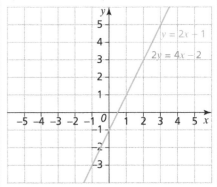

Infinite solutions (the same line)

Linear and quadratic simultaneous equations have zero, one or two points of intersection

$y = (x - 2)^2 + 1.5$ and $y = \frac{2}{3}x - 2$

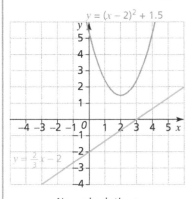

No real solutions

$y = (x - 2)^2 + 1.5$ and $y = 2x - 3.5$

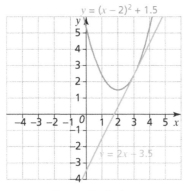

One real solution

$y = (x - 2)^2 + 1.5$ and $y = \frac{2}{3}x + 1$

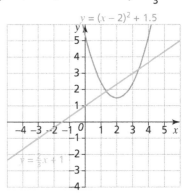

Two real solutions

Quick Test

1. Solve the simultaneous equations $y - 2x = -5$ and $3x - 4y = 10$.
2. Solve the simultaneous equations $2x + 3y = -3$ and $3x = -4y - 5$.
3. Solve the simultaneous equations $2x + y = 4$ and $y = x^2 + 2x - 8$.

Key Words

simultaneous equations
elimination method
substitution method

Inequalities

You must be able to:

- Solve linear and quadratic inequalities with one variable
- Express solutions using set notation
- Represent inequalities graphically.

Set Notation

- A **set** is a collection of elements and is denoted using curly brackets, { }
- The notation {x: some criteria} shows the set of values for which x satisfies some criteria.
- $x < a$ can be written {x: $x < a$} and, similarly, $x > b$ can be written {x: $x > b$}
- 'And' is written with the symbol $\cap$ and 'Or' is written using the symbol $\cup$
- For example, $x < 1$ or $x > 3$ is written {x: $x < 1$}$\cup${x: $x > 3$}

- $x \geqslant -3$ and $x < 2$ is written {x: $x \geqslant -3$}$\cap$ {x: $x < 2$}, or as {x: $-3 \leqslant x < 2$}

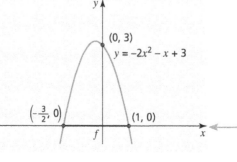

> ### Key Point
> $\cap$ means 'and'.
> $\cup$ means 'or'.

> ### Key Point
> A number line can help to visualise the values for which the inequalities hold. Use a solid circle for $\geqslant$ and $\leqslant$, but use an open circle for $<$ and $>$.

Solving Linear Inequalities Algebraically

- To solve an **inequality** means to find the values for which the inequality holds true.

Find the set of values for which $5 - 2x > -15$ and $4(x + 4) \geqslant 18$.

$$5 - 2x > -15 \quad \text{and} \quad 4(x + 4) \geqslant 18$$
$$-2x > -20 \qquad\qquad 4x + 16 \geqslant 18$$
$$x < 10 \qquad\qquad 4x \geqslant 2, \text{ so } x \geqslant \tfrac{1}{2}$$

The solutions can be written in set notation as $\left\{x: x \geqslant \dfrac{1}{2}\right\} \cap \{x: x < 10\}$,

or on a number line or as a combined inequality $\tfrac{1}{2} \leqslant x < 10$.

> ### Key Point
> Solve inequalities just like equations, but don't forget to reverse the inequality symbol when multiplying or dividing by a negative number.

Solving Quadratic Inequalities

- Sketch the quadratic and determine the range of x-values for which the inequality holds.

1) Solve $-2x^2 - x \geqslant -3$.

$$-2x^2 - x + 3 \geqslant 0$$

> Sketch the graph. The function is $\geqslant 0$ when it is at or above the x-axis.

$-\dfrac{3}{2} \leqslant x \leqslant 1$, or using set notation

$\left\{x: -\dfrac{3}{2} \leqslant x\right\} \cap \{x: x \leqslant 1\}$.

See page 13 for graphing quadratics.

2) Solve $\frac{3}{x} \le 5$, $x \ne 0$.

> Multiply each side by x^2 because x could be positive or negative.

$$3x \le 5x^2$$
$$3x - 5x^2 \le 0$$
$$x(3 - 5x) \le 0$$
$$x(3 - 5x) = 0, \; x = 0 \text{ or } x = \tfrac{3}{5}$$

The values of x for which

$y \le 0$ are either $x < 0$ or $x \ge \tfrac{3}{5}$,

$\{x: x < 0\} \cup \{x: x \ge \tfrac{3}{5}\}$

> Solve the quadratic equation to find the x-intercepts and sketch the graph.

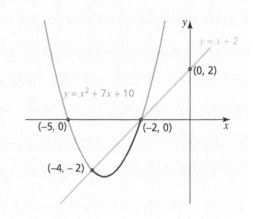

$y = -5x^2 + 3x$
$(0, 0)$
$\left(\tfrac{3}{5}, 0\right)$

> **Key Point**
>
> To solve a quadratic inequality, sketch the corresponding quadratic equation – only the x-intercepts and a general correct shape are needed.

Simultaneous Inequalities

- Inequalities can be given in the form f(x) > g(x) given the functions f(x) and g(x).
- The values of x for which f(x) < g(x) are the x-values where the curve f(x) is below the curve g(x).
- The values of x for which f(x) > g(x) are the x-values where the curve f(x) is above the curve g(x).

f(x) = $x^2 + 7x + 10$, g(x) = $x + 2$

Find the values of x for which f(x) < g(x).

> Sketch the graphs of y = f(x) and y = g(x) and find the points of intersection.

$$x^2 + 7x + 10 = x + 2$$
$$x^2 + 6x + 8 = 0$$
$$(x + 2)(x + 4) = 0$$
$$x = -2, \; x = -4$$

> Substitute the values of x to find the points of intersection $(-2, 0)$ and $(-4, -2)$.

The values of x for which the curve of f(x) is below the line g(x) are $-4 < x < -2$.

$y = x + 2$
$(0, 2)$
$y = x^2 + 7x + 10$
$(-5, 0)$
$(-2, 0)$
$(-4, -2)$

Representing Inequalities Graphically

Draw a graph to show the region of values satisfied by the inequalities $y < \tfrac{1}{2}x + 1$ and $y \ge -x + 1$.

> Graph both functions and shade the area that satisfies both inequalities.

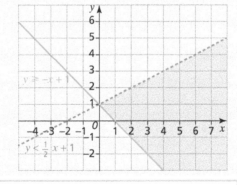

$y \ge -x + 1$
$y < \tfrac{1}{2}x + 1$

> **Key Point**
>
> Use a dotted line or curve for the symbols > or <.
>
> Use a solid line or curve for the symbols $\ge$ or $\le$.
>
> Shade the areas satisfied by the inequalities.

> **Quick Test**
>
> 1. Solve the inequality $3(x - 4) + 2 \le 11$.
> 2. Find the values of x for which $\frac{-2}{x} \ge 8$.
> 3. Solve the inequality $2x^2 - 5x - 3 > 5x - 11$.

> **Key Words**
>
> set
> inequality

Sketching Curves

You must be able to:

- Graph simple cubic, quartic and functions of the form $y = \frac{a}{x}$ and $y = \frac{a}{x^2}$
- Interpret algebraic solutions graphically
- Use and graph direct proportion to show a relationship between two variables.

General Shape of Graphs

- The sign of 'a' ($a > 0$ or $a < 0$) affects the shape of the graph:

$a > 0$

$a < 0$

| Cubic | Quartic | $y = \frac{a}{x}$ | $y = \frac{a}{x^2}$ | Cubic | Quartic | $y = \frac{a}{x}$ | $y = \frac{a}{x^2}$ |

Cubic and Quartic Functions

- The roots of any polynomial function are the x-intercepts.
- A polynomial of power n will have at most n real roots.

> **Key Point**
>
> A **cubic** function has the form $y = ax^3 + bx^2 + cx + d$ and a **quartic** function has the form $y = ax^4 + bx^3 + cx^2 + dx + e$.

Sketch a graph of the function $f(x) = x^3 - 5x^2 + 2x + 8$ given that $x = 2$ is a root of the function.

Dividing: $(x^3 - 5x^2 + 2x + 8) \div (x - 2) = x^2 - 3x - 4$

Factorising: $x^2 - 3x - 4 = (x - 4)(x + 1)$

$f(x) = x^3 - 5x^2 + 2x + 8 = (x - 2)(x - 4)(x + 1)$

The function intersects the x-axis at $(2, 0)$, $(4, 0)$ and $(-1, 0)$ and intersects the y-axis at $(0, 8)$.

If $x = 2$ is a root, then $(x - 2)$ is a factor of $f(x)$. See page 11 for polynomial division.

The x-intercepts are the solutions to the equation $f(x) = 0$ and the y-intercept is the value of $f(0)$.

See pages 8–9 for factorisation.

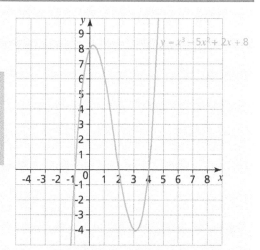

Functions $y = \frac{a}{x}$ and $y = \frac{a}{x^2}$

- The graphs of $y = \frac{a}{x + b} + c$ and $y = \frac{a}{(x + b)^2} + c$ have **asymptotes** at $y = c$ and $x = -b$.

Sketch the graph of the equation $y = \frac{2}{(x - 1)^2} - 2$.

$y = \frac{2}{(x - 1)^2} - 2$ has asymptotes at $y = -2$ and $x = 1$.

x	−1	0	1	2	3
y	−1.5	0	Undefined	0	−1.5

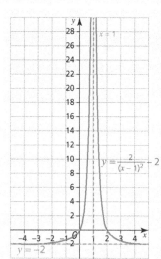

Identify the asymptotes by considering values of x and y for which the equation is undefined.

Sketch the curve remembering the general shape of the graph and the asymptotes.

A graph of $y = \frac{a}{(x + b)^2} + c$ is a translation of the graph $y = \frac{a}{x^2}$ by the vector $\begin{pmatrix} -b \\ c \end{pmatrix}$. See pages 20–23 for graph transformations.

Calculate a few points and plot the general shape of the curve, remembering the asymptotes.

Interpreting Algebraic Equations

- A graph can be used to solve an equation or a set of simultaneous equations.

Sketch a graph to find the number of real solutions to the equation $(x + 1)(x - 2)^2(x + 3) = x - 4$ and estimate the solutions.

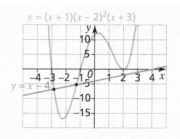

The curve and the line intersect at two points so there are two real solutions to the equation; $x \approx -2.8$ and $x \approx -1.3$

Direct Proportion

- Two variables are in **direct proportion** when an increase or decrease in one affects the other at the same rate.
- A graph of a direct proportion will be a straight line through the origin. The symbol $\propto$ means 'in proportion'.

Key Point

$y \propto x$ if $y = kx$ for some constant k.

The cost to hire bowling shoes in pounds, c, is directly proportional to the number of games played, n. To hire shoes for four games costs £19.00.

Write an equation and draw a graph expressing the proportional relationship between c and n.

$c \propto n$, so $c = kn$

$19 = 4k$, $k = 4.75$ ⟵ Substitute the values and solve for k.

$c = 4.75n$ ⟵ Rewrite the equation using the value of k you found.

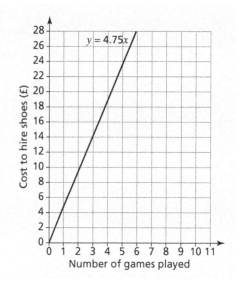

Quick Test

1. Sketch a graph of $f(x) = x(x + 1)(x + 2)(x - 2)$.
2. Sketch the graph of $y = \dfrac{3}{x - 1}$
3. Identify which of the graphs (right) represent direct proportion and write down the value of the constant of proportionality.

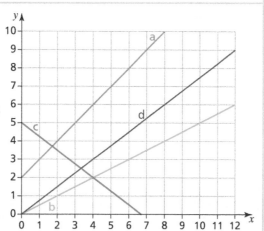

Key Point

The x-value(s) at the point(s) of intersection of $f(x)$ and $g(x)$ are the solutions to the equation $f(x) = g(x)$.

Key Words

cubic
quartic
asymptote
direct proportion

Translating Graphs

You must be able to:

- Understand the effect of a transformation $y = f(x + a)$ and $y = f(x) + a$ on a given function $y = f(x)$
- Sketch the graph resulting from a transformation $y = f(x + a)$ and $y = f(x) + a$ on a given function $y = f(x)$.

Transformations of the Form $y = f(x + a)$

- A **transformation** of the form $y = f(x + a)$ is a **translation** of the graph $y = f(x)$ by $-a$ in the x-direction, i.e. a translation by the **vector** $\begin{pmatrix} -a \\ 0 \end{pmatrix}$.

Key Point

$y = f(x + a)$ is a translation of $y = f(x)$ by the vector $\begin{pmatrix} -a \\ 0 \end{pmatrix}$.

1) Given the graph of $f(x) = -\dfrac{5}{x^2}$, draw the graph of $y = -\dfrac{5}{(x + 1)^2}$ and identify the asymptotes of each function.

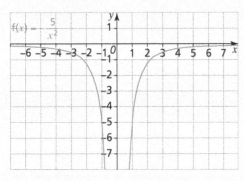

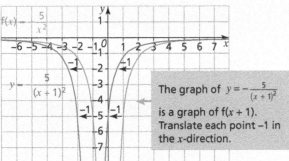

The graph of $y = -\dfrac{5}{(x + 1)^2}$ is a graph of $f(x + 1)$. Translate each point -1 in the x-direction.

The asymptotes of $f(x) = -\dfrac{5}{x^2}$ are

$x = 0$ and $y = 0$.

The asymptotes of $f(x + 1)$ are

$x = -1$ and $y = 0$. ← When a function is translated in either the x- or y-direction, any asymptotes are also translated.

2) Given $f(x) = (x + 2)(x + 0.5)(x - 1)(x - 2)$, write down the equation of $y = f(x - 2)$ and draw the graph $y = f(x - 2)$.

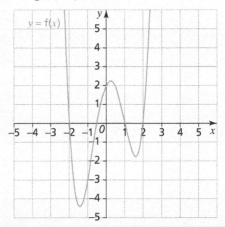

To find the equation of $y = f(x - 2)$, substitute $x - 2$ in for x in $f(x)$.

$f(x - 2) = (x - 2 + 2)(x - 2 + 0.5)(x - 2 - 1)(x - 2 - 2)$

$\qquad\quad = x(x - 1.5)(x - 3)(x - 4)$

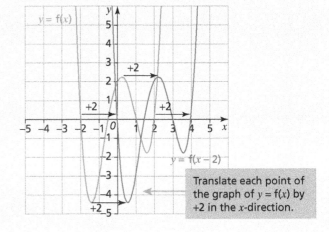

Translate each point of the graph of $y = f(x)$ by $+2$ in the x-direction.

Transformations of the Form $y = f(x) + a$

- A transformation of the form $y = f(x) + a$ is a translation of the graph $y = f(x)$ by a in the y-direction, i.e. a translation by the vector $\begin{pmatrix} 0 \\ a \end{pmatrix}$.

> **Key Point**
>
> $y = f(x) + a$ is a translation of $y = f(x)$ by the vector $\begin{pmatrix} 0 \\ a \end{pmatrix}$.

1) Given the graph of $f(x) = -3x^2 - 2x + 4$, sketch the graph of $g(x) = -3x^2 - 2x + 1$.

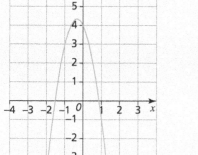

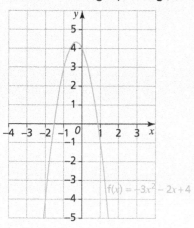

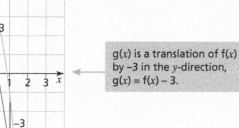

g(x) is a translation of f(x) by −3 in the y-direction, $g(x) = f(x) − 3$.

2) Given the graph of $y = f(x)$, sketch the graph of $y = f(x) + 2$.

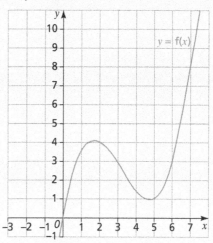

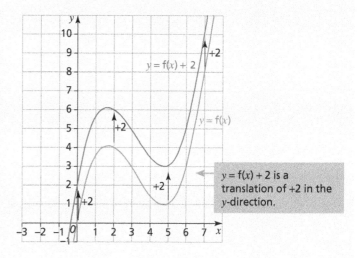

$y = f(x) + 2$ is a translation of +2 in the y-direction.

Quick Test

1. Given the function
   $f(x) = (x - 1)(x - 2)(x + 1)(x + 3)$
   a) Find the equation of the curve $y = f(x + 1)$.
   b) Find the equation of the curve $y = f(x) + 1$.
2. Given the graph of $y = f(x)$, shown right:
   a) Sketch the graph of $y = f(x - 2)$.
   b) Sketch the graph of $y = f(x) - 2$.

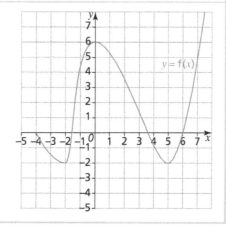

> **Key Words**
>
> transformation
> translation
> vector

Stretching and Reflecting Graphs

You must be able to:

- Understand the effect of a transformation $y = af(x)$, $y = -f(x)$, $y = f(ax)$ and $y = f(-x)$ on a given function $y = f(x)$
- Sketch the graph resulting from a transformation $y = af(x)$ and $y = f(ax)$ on a given function $y = f(x)$
- Sketch the reflection resulting from a transformation $y = -f(x)$ and $y = f(-x)$ on a given function $y = f(x)$.

Transformations of the Form $y = af(x)$

- A transformation of the form $y = af(x)$ is a stretch of the graph $y = f(x)$ by a **scale factor** of a in the y-direction.
- Each point (x, y) in $y = f(x)$ maps on to the point (x, ay) in $y = af(x)$.

> **Key Point**
>
> $y = af(x)$ stretches $y = f(x)$ by a scale factor of a vertically.

Transformations of the Form $y = f(ax)$

- A transformation of the form $y = f(ax)$ is a stretch of the graph $y = f(x)$ by a factor of $\frac{1}{a}$ in the x-direction.
- Each point (x, y) in $y = f(x)$ maps on to the point $\left(\frac{1}{a}x, y\right)$ in $y = f(ax)$.

> **Key Point**
>
> $y = f(ax)$ stretches $y = f(x)$ by a factor of $\frac{1}{a}$ horizontally ($a \neq 0$).

Given the graph of $y = f(x)$, shown right, sketch the graphs of $y = 2f(x)$ and $y = f(2x)$.

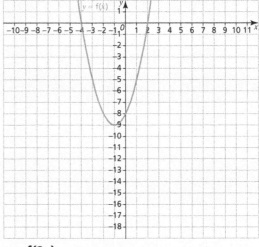

$y = 2f(x)$

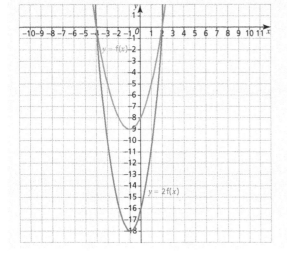

$y = f(2x)$

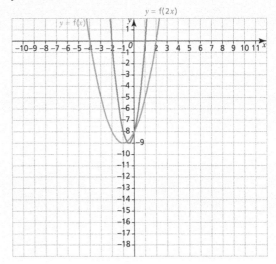

Reflections in the x- and y-axes

- A transformation $y = f(-x)$ is a **reflection** in the y-axis and maps each point (x, y) in $y = f(x)$ on to $(-x, y)$ in $y = f(-x)$.
- A transformation $y = -f(x)$ is a reflection in the x-axis and maps each point (x, y) in $y = f(x)$ on to $(x, -y)$ in $y = -f(x)$.

Given the graph of $y = f(x)$, sketch the graphs of $y = f(-x)$ and $y = -f(x)$.

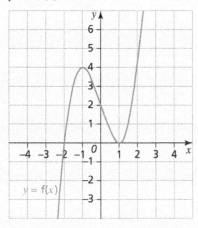

Key Point
$y = f(-x)$ is a reflection of the graph $y = f(x)$ in the y-axis. $y = -f(x)$ is a reflection of the graph $y = f(x)$ in the x-axis.

To sketch **y = f(–x)**, reflect the graph $y = f(x)$ in the y-axis.

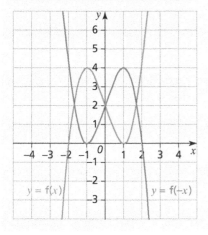

To sketch **y = –f(x)**, reflect the graph $y = f(x)$ in the x-axis.

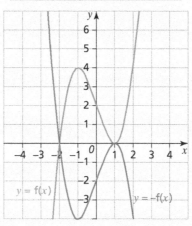

Quick Test

Given the graph of $y = f(x)$, shown right:

1. Sketch the graph of $y = \frac{1}{2} f(x)$.

2. Sketch the graph of $y = f\left(\frac{1}{2} x\right)$.

3. Sketch the graph of $y = -f(x)$.

4. Sketch the graph of $y = f(-x)$.

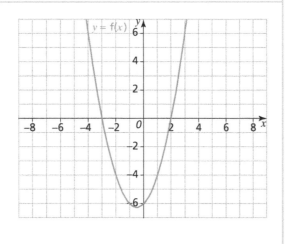

Key Words
scale factor reflection

Indices and Surds

1 Simplify the following:

a) $2x^2y^3 \times 3x^4y^5$ [1]

b) $\left(3m^4n^{\frac{3}{2}}\right)^4$ [1]

c) $\dfrac{\left(16k^6j^2\right)^{\frac{1}{2}}}{4j}$ [2]

2 Simplify the following:

a) $\left(3+\sqrt{2}\right)-\left(1-\sqrt{8}\right)$ [1]

b) $\left(2+\sqrt{7}\right)\left(3+\sqrt{7}\right)$ [1]

3 Rationalise the denominators of the following:

a) $\dfrac{2}{4+\sqrt{7}}$ [2]

b) $\dfrac{3+\sqrt{10}}{3-\sqrt{10}}$ [2]

> **Total Marks** .... / 10

Manipulating Algebraic Expressions

1 Factorise completely $7k^2 - 22k + 3$. [1]

2 Expand and simplify $(x + 2)(3x - 8) - 2x(y + 4) - 3y$. [2]

3 Write the area of the trapezium as both a factorised and expanded expression. [3]

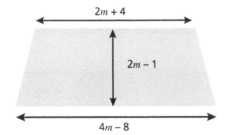

2m + 4

2m − 1

4m − 8

> **Total Marks** .... / 6

Expanding and Dividing Polynomials

1. The coefficient of x^4 in the expansion of $(3 - bx)^5$ is 240.

   Find the possible values of b. [3]

2. $f(x) = ax^3 + x^2 - 13x + 6$

   Given that $(x + 3)$ is a factor of f(x), find the value of a. [3]

   Total Marks .... / 6

Quadratic Equations

1. Show that any quadratic equation $ax^2 + bx + c = 0$
   can be written in the form $p(x + q)^2 + r = 0$. [3]

2. Graph f(x) = $2x^2 + 2x - 4$. [5]

3. A child's swing is pulled back then released. The motion of the arc created when the swing is released can be modelled using a quadratic function of the form f(x) = $2x^2 + kx + c$, for some value k and c.

   Given that the swing is at its lowest point of 0.6 m after one second, find the height from which the swing was released. [4]

   Total Marks .... / 12

Simultaneous Equations

1 Solve the simultaneous equations $4x + 2y = 2$ and $y = 2x^2 - x$. [4]

2 Solve the simultaneous equations $y = x - 1$ and $2x^2 - xy - 3y^2 = 1$. [5]

Total Marks _____ / 9

Inequalities

1 Solve $6x^2 + 7x < 3$, giving your answer in set notation. [4]

2 Find the values of x for which $-3(x - 1) + 4 > 5$. [2]

3 Find the values of x for which $-\dfrac{3}{x} \geqslant 2$. [4]

Total Marks _____ / 10

Sketching Curves

1 Sketch a graph of $f(x) = x^4 + x^3 - 2x^2$, given that $x = 1$ is a root of $f(x)$. [4]

2 Sketch a graph to show the solution(s) to the equation $2x^3 - x^2 - x = \dfrac{2}{x^2}$ [5]

3 Write down the equation of the graph shown below, given it is of the form $y = (x + a)(x + b)(x + c)$, and find the value of the y-intercept. [2]

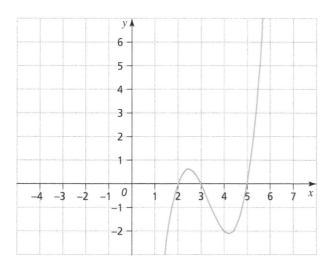

 Total Marks _____ / 11

Practice Questions

Translating Graphs

1 Given $f(x) = 2x^2 - 3x - 2$, find the roots of $f(x + 2)$. [3]

2 Given $f(x) = \dfrac{3}{x^2}$

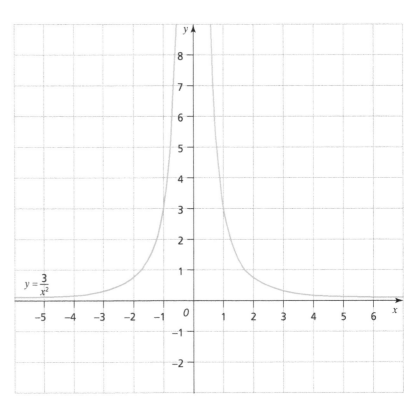

$y = \dfrac{3}{x^2}$

a) Sketch the graph of $y = f(x) - 2$ on the grid above, noting the asymptotes. [2]

b) Sketch the graph of $y = f(x - 2)$ on the grid above, noting the asymptotes. [2]

Total Marks _____ / 7

Stretching and Reflecting Graphs

1 The point A(3, −20) lies on the curve $y = f(x)$.

Write down the coordinates of point A after the transformation:

a) $y = f\left(\frac{1}{5}x\right)$ [1]

b) $y = 5f(x)$ [1]

2 Given the graph of $y = f(x)$ shown right, sketch the graph of $y = f(−x)$, stating clearly the effect on the asymptotes. [3]

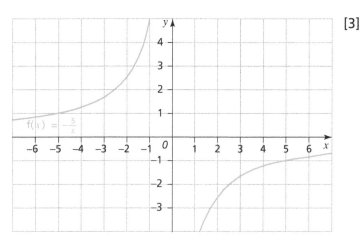

$f(x) = -\frac{5}{x}$

3 Given the graph of $f(x)$ below, sketch the graph of $−2y = f(x)$. [4]

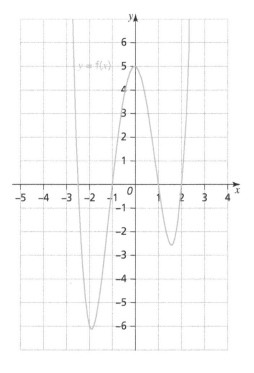

$y = f(x)$

Total Marks _____ / 9

Equations of Straight Lines

You must be able to:

- Use the equation of a straight line in the forms $y = mx + c$, $y - y_1 = m(x - x_1)$ and $ax + by + c = 0$
- Recall and use the relationship between gradients of both parallel and perpendicular lines
- Use straight lines to solve questions in context.

Finding Equations, Midpoints and Distance

- Equations of straight lines can be written in the forms:
  - $y = mx + c$, where m is the **gradient** and c is the **y-intercept**
  - $y - y_1 = m(x - x_1)$, where m is gradient and (x_1, y_1) is a point on the line
  - $ax + by + c = 0$, where $-\dfrac{a}{b}$ is the gradient and $-\dfrac{c}{b}$ is the y-intercept.

Find the equation of the line that passes through the points $(-2, 3)$ and $(3, 6)$. Express your answer in the forms:

a) $y = mx + c$

$m = \dfrac{y_2 - y_1}{x_2 - x_1} = \dfrac{6 - 3}{3 - (-2)} = \dfrac{3}{5}$

$y - y_1 = m(x - x_1)$

$y - 3 = \dfrac{3}{5}(x - (-2))$ ⟵ Substitute **either** given point and the gradient.

$y - 3 = \dfrac{3}{5}(x + 2)$

$y = \dfrac{3}{5}x + \dfrac{21}{5}$

b) $ax + by + c = 0$

> Multiply each term from **a)** by 5.

$5y = 3x + 21$

$-3x + 5y - 21 = 0$

> **Key Point**
>
> Remember the gradient is $m = \dfrac{y_2 - y_1}{x_2 - x_1}$ and watch your signs when substituting in the values.

> **Key Point**
>
> To find an equation of a line passing through a point, substitute the gradient and a point on the line into either $y = mx + c$ or $y - y_1 = m(x - x_1)$.

- The **midpoint** between two points (x_1, y_1) and (x_2, y_2) is $\left(\dfrac{x_1 + x_2}{2}, \dfrac{y_1 + y_2}{2}\right)$.

- The distance between two points (x_1, y_1) and (x_2, y_2) can be calculated using the (Pythagoras) formula $d = \sqrt{(x_2 - x_1)^2 + (y_2 - y_1)^2}$.

Find the midpoint and the distance between the points $(3, 5)$ and $(8, 4)$.

Midpoint: $\left(\dfrac{3+8}{2}, \dfrac{5+4}{2}\right) = \left(\dfrac{11}{2}, \dfrac{9}{2}\right)$ or $(5.5, 4.5)$

$d = \sqrt{(8-3)^2 + (4-5)^2} = \sqrt{26} \approx 5.10$ units

> **Key Point**
>
> To find the gradient of a perpendicular line, just flip the fraction and change the sign.

- **Parallel** lines have the same gradient; $m_1 = m_2$.
- The product of the gradients of **perpendicular** lines is -1; $m_1 \times m_2 = -1$.

  1) Line l is perpendicular to the line k, $4x + 3y - 9 = 0$, and passes through the point $(4, 2)$. Find the equation of line l.

  The gradient of line k is $-\dfrac{4}{3}$ so the gradient of line l is $\dfrac{3}{4}$. ⟵

  > Rearrange $4x + 3y - 9 = 0$ to find the gradient: $y = -\dfrac{4}{3}x + 9$.

  $y - y_1 = m(x - x_1)$

  $y - 2 = \dfrac{3}{4}(x - 4)$, so $y - 2 = \dfrac{3}{4}x - 3$

  $y = \dfrac{3}{4}x - 1$ ⟵

  > Do not leave your answer in the form $y - y_1 = m(x - x_1)$. Answers should be simplified to earn full marks.

2) Find the equation of the line parallel to $-3x + 5y - 10 = 0$ that passes through the point (5, 6).

Substituting the given gradient and point into $y = mx + c$ gives

$6 = \left(\dfrac{3}{5} \times 5\right) + c$, so $c = 3$ ← The gradient of the line $-3x + 5y - 10 = 0$ is $\dfrac{3}{5}$.

$y = \dfrac{3}{5}x + 3$

Solving Problems in Context

1) A house painter charges a fixed call-out fee and an additional daily rate. House A takes four days to paint and she charges £850. House B takes seven days to paint and she charges £1300.

a) Find an equation expressing the total cost, C, in terms of the number of days of work, d.

$m = \dfrac{1300 - 850}{7 - 4} = \dfrac{450}{3} = 150$ ← Use the points (4, 850) and (7, 1300) to find the gradient.

$850 = (150 \times 4) + c$ ← Substitute into $y = mx + c$.

$c = 250$

$C = 150d + 250$

b) She charges £550 to paint a third house. Find the number of days it took her to paint the house.

$550 = 150d + 250$ ← Substitute into $C = 150d + 250$.

$300 = 150d$

$d = 2$, so it took her two days to paint the house.

2) The table (right) shows the highest temperatures in London in degrees Celsius and degrees Fahrenheit over a one-week period.

Write an equation expressing the temperature in °F in terms of the temperature in °C.

$m = \dfrac{y_2 - y_1}{x_2 - x_1} = \dfrac{75.2 - 69.8}{24 - 21} = \dfrac{9}{5}$ ← Using the points (21, 69.8) and (24, 75.2), calculate the gradient and then use $(y - y_1) = m(x - x_1)$.

$y - 69.8 = \dfrac{9}{5}(x - 21)$, so $y = \dfrac{9}{5}x + 32$

$°F = \dfrac{9}{5}°C + 32$

Day	°C	°F
Mon	21	69.8
Tue	24	75.2
Wed	27	80.6
Thu	28	82.4
Fri	26	78.8
Sat	25	77
Sun	23	73.4

Quick Test

1. Find the equation of the line between the points (−3, 3) and (9, −5).
2. Find the midpoint and the distance between points (3, 8) and (6, 12).
3. Find the equation of the line parallel to $y = \dfrac{2}{5}x - 7$ which passes through the point (5, −1). Give your answer in the form $ax + by + c = 0$.
4. Write an equation of a line that is perpendicular to $4x + 5y + 2 = 0$.

Key Words

gradient
y-intercept
midpoint
parallel
perpendicular

Circles

You must be able to:

- Understand the equation of a circle $(x - a)^2 + (y - b)^2 = r^2$
- Complete the square to find the centre and radius of a circle
- Use the properties of circles to find the equation of a circle and the equation of a tangent at a point on a circle
- Understand and use the properties of circumcircles to find the equation of a circle.

The Equation and Centre of a Circle

- $(x - a)^2 + (y - b)^2 = r^2$, where (a, b) is the centre and r is the **radius**.
- Complete the square to find the centre and the radius of a circle with an equation given in the form $x^2 + y^2 + 2fx + 2gy + c = 0$.

> **Key Point**
>
> The general equation of a circle is $(x - a)^2 + (y - b)^2 = r^2$.

Find the centre and the radius of the circle with equation $x^2 + 2x + y^2 - 10y + 10 = 0$.

> Complete the square on both the x and y terms.

$x^2 + 2x + y^2 - 10y = -10$

$(x + 1)^2 + (y - 5)^2 = -10 + 1 + 25$ ← See pages 12–13 for completing the square.

$(x + 1)^2 + (y - 5)^2 = 16$

The centre is $(-1, 5)$ and the radius is 4.

Using Circle Theorems

Properties of a Circle

 The angle at the circumference of a semicircle is a right angle.

The perpendicular from the centre to the chord bisects the chord.

 The angle between a tangent and a radius is 90 degrees.

Show that AB is a **diameter** of a circle that passes through points A(–1, 1), B(5, 9) and C(2, 10). Find the equation of the circle.

> Start with a sketch to help you visualise the problem.

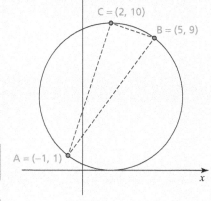

$m_{AC} = \dfrac{y_2 - y_1}{x_2 - x_1} = \dfrac{10 - 1}{2 - (-1)} = 3$ 　　　$m_{BC} = \dfrac{10 - 9}{2 - 5} = -\dfrac{1}{3}$

$m_{AC} \times m_{BC} = 3 \times \left(-\dfrac{1}{3}\right) = -1$, so AC and BC are perpendicular, therefore

angle ACB is the angle in a semicircle and AB must be a diameter.

$\left(\dfrac{x_1 + x_2}{2}, \dfrac{y_1 + y_2}{2}\right) = \left(\dfrac{-1 + 5}{2}, \dfrac{1 + 9}{2}\right) = (2, 5)$ ← The midpoint of the diameter is the centre of the circle.

$r = \sqrt{(5 - 2)^2 + (9 - 5)^2} = 5$ ← Substitute $(2, 5)$ and $(5, 9)$ into the distance formula.

The equation is $(x - 2)^2 + (y - 5)^2 = 25$.

Finding the Equation of the Tangent

- A **tangent** just touches the circle at one point. The tangent is perpendicular to the radius at that point.

Point A(9, 7) lies on the circle $(x - 3)^2 + (y + 1)^2 = 100$. Find the equation of the tangent line at A.

The gradient of the radius at A is $m = \dfrac{x_2 - x_1}{y_2 - y_1} = \dfrac{7 - (-1)}{9 - 3} = \dfrac{4}{3}$ so the gradient of the tangent is $-\dfrac{3}{4}$.

$7 = \left(-\dfrac{3}{4} \times 9 \right) + c$ ← Substitute the coordinates of A and the gradient into $y = mx + c$.

$c = \dfrac{55}{4}$

$y = -\dfrac{3}{4}x + \dfrac{55}{4}$

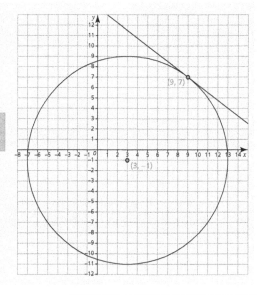

Circumcircles

- A unique circle can be drawn through any three points that form the vertices of a triangle. This is called a **circumcircle**.
- The perpendicular bisectors of each side of the **circumscribed triangle** intersect at the centre of the circle, also called the **circumcentre**.

A(–3, 4), B(5, 8) and C(2, –1) form the vertices of a circumscribed triangle. Find the equation of the circumcircle.

The intersection of the perpendicular bisectors is the circumcentre.

The perpendicular bisector to AB is $y = -2x + 8$.

The perpendicular bisector to AC is $y = x + 2$. ←

The perpendicular bisectors intersect at (2, 4). ←

$r = \sqrt{(2 - (-3))^2 + (4 - 4)^2} = 5$ ←

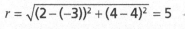
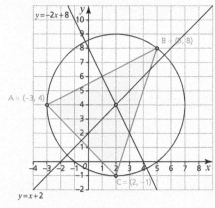

Substitute the centre and a point on the circumference.

The equation of the circumcircle is $(x - 2)^2 + (y - 4)^2 = 25$.

> **Key Point**
>
> To show a point is on a line, a curve or a circle, substitute that point into the equation and show that the equation still holds.

Find the equation of the perpendicular bisectors by finding the midpoint and the gradient between the two points and substituting into $y - y_1 = m(x - x_1)$.

Use simultaneous equations.

> **Key Words**
>
> radius
> diameter
> tangent
> circumcircle
> circumscribed triangle
> circumcentre

Quick Test

1. Find the centre and radius of the circle $x^2 + y^2 - 10x - 2y + 14 = 0$.
2. Find the equation of the tangent line to the circle $x^2 + y^2 + 6x + 4y = 87$ at the point (5, 4).

Sine, Cosine & Tangent Functions, and Sine & Cosine Rule

You must be able to:

- Understand and use the sine, cosine and tangent functions; their graphs, symmetries and periodicity
- Understand and use the definitions of sine, cosine and tangent for all arguments
- Apply the sine and cosine rules and find the area of a triangle using $\frac{1}{2}ab\sin C$.

Basic Trigonometric Functions and their Graphs

- You can use the following ratios to find missing sides and angles in a right-angle triangle, i.e. $0° < \theta \leqslant 90°$.

$$\sin\theta = \frac{\text{opposite}}{\text{hypotenuse}} \qquad \cos\theta = \frac{\text{adjacent}}{\text{hypotenuse}} \qquad \tan\theta = \frac{\text{opposite}}{\text{adjacent}}$$

- Our understanding of trigonometric functions can be extended to angles outside the interval $0° < \theta \leqslant 90°$.
- Sine, cosine and tangent functions repeat themselves after a certain interval.
- The interval is called the **period** of the function.
- Graphs of trigonometric functions have many symmetrical properties:

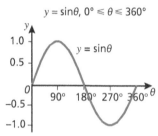
$y = \sin\theta, 0° \leqslant \theta \leqslant 360°$

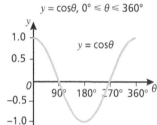
$y = \cos\theta, 0° \leqslant \theta \leqslant 360°$

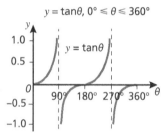
$y = \tan\theta, 0° \leqslant \theta \leqslant 360°$

> ### Key Point
>
> $\sin\theta$ has a period of 360° or 2π.
>
> $\cos\theta$ has a period of 360° or 2π.
>
> $\tan\theta$ has a period of 180° or π.
>
> $\sin\theta$ has **symmetry** about $\theta = 90°$, $0° \leqslant \theta \leqslant 180°$.
>
> $\cos\theta$ has symmetry about $\theta = 0°$.
>
> $\tan\theta$ has **asymptotes** at $\theta = (2n + 1)90°$, where n is an integer.
>
> $\cos\theta = \sin(90° + \theta)$.

Using the Functions for all Arguments

- You must be able to understand and use the definitions of $\sin x$, $\cos x$ and $\tan x$ for all **arguments**.
- The x-y plane is divided into quadrants:
  - in the first quadrant (0° to 90°) $\sin x$, $\cos x$ and $\tan x$ are all positive
  - in the second quadrant (90° to 180°) only $\sin x$ is positive
  - in the third quadrant (180° to 270°) only $\tan x$ is positive
  - in the fourth quadrant (270° to 360°) only $\cos x$ is positive.
- Angles which lie outside the range 0° to 360° will still lie in one of the four quadrants owing to the cyclic nature of trigonometric functions.

CAST Diagram

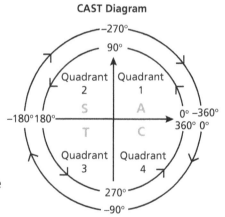

Express, in terms of trigonometric ratios of acute angles, $\sin(300)°$.

Method 1

The acute angle made with the x-axis is 60°.

In the fourth quadrant only $\cos x$ is positive, so $\sin x$ is negative.

$\sin(300)° = -\sin(60)°$

Method 2

You can also use the symmetry of the sine graph:

$\sin(300)° = \sin(240)° = -\sin(120)°$

So $\sin(300)° = -\sin(60)°$

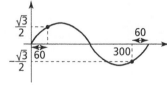

Graph Transformations

- The rules covered on pages 20–21 of this revision guide apply to trigonometric functions.

Sketch the graph of $y = 2 + \sin 2\theta$, $0° \leqslant \theta \leqslant 360°$.

If $f(\theta) = \sin\theta$, $2 + \sin 2\theta = f(2\theta) + 2$.

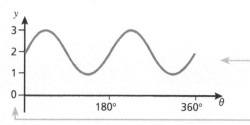

A stretch of scale factor $\frac{1}{2}$ in the direction of the x-axis.

A translation of $+2$ in the direction of the y-axis.

Sine and Cosine Rule and Area of a Triangle

$\frac{a}{\sin A} = \frac{b}{\sin B} = \frac{c}{\sin C}$	$a^2 = b^2 + c^2 - 2bc \cos A$	$A = \frac{1}{2}ab \sin C$

1) Find the value of x.

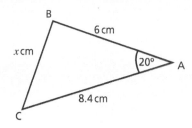

$a^2 = b^2 + c^2 - 2bc \cos A$

$a = x$, $b = 8.4$, $c = 6$, $A = 20$

$x^2 = 8.4^2 + 6^2 - 2 \times 8.4 \times 6 \times \cos 20$

$x^2 = 11.83898382$

$x = 3.44$ (3 s.f.)

2) Find the area of this triangle.

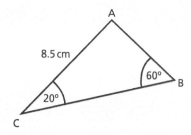

$\frac{a}{\sin A} = \frac{b}{\sin B}$

$b = 8.5 \qquad A = 100 \qquad B = 60$

$\frac{a}{\sin 100} = \frac{8.5}{\sin 60}$

$a = \frac{8.5 \sin 100}{\sin 60}$

$a = 9.665843362$

$\text{Area} = \frac{1}{2}ab \sin C$

$a = 9.665843362 \qquad b = 8.5 \qquad C = 20$

$\text{Area} = \frac{1}{2} \times 9.665843362 \times 8.5 \times \sin 20$

$\text{Area} = 14.1 \text{ cm}^2$ (3 s.f.)

> **Key Point**
>
> When labelling the triangle, ensure side a is opposite angle A, etc.

> **Key Point**
>
> Use the cosine rule when you are given either **two sides and the angle between them** or **three sides**.
>
> For all other combinations use the sine rule.

To find the area, first find the length of side a.

Note the missing angle is 100°.

> **Key Point**
>
> Remember to check your calculator is in the correct mode.

Quick Test

1. Sketch the graph $y = \cos(2\theta - 30)$, $0 \leqslant \theta \leqslant 360°$.
2. Express, in terms of trigonometric ratios of acute angles, $\cos(400°)$.
3. In triangle ABC, AB $= 3.8$ cm, BC $= 5.2$ cm and angle BAC $= 35°$. Find angle ABC.
4. In triangle DEF, DE $= 5$ cm, EF $= 6.2$ cm and DF $= 10$ cm. Find the value of the smallest angle.

> **Key Words**
>
> period
> symmetry
> asymptote
> argument

Trigonometric Identities

You must be able to:

- Understand and use $\tan\theta \equiv \dfrac{\sin\theta}{\cos\theta}$
- Understand and use $\sin^2\theta + \cos^2\theta = 1$
- Use these identities to prove further identities.

$\tan\theta \equiv \dfrac{\sin\theta}{\cos\theta}$

- $\tan\theta \equiv \dfrac{\sin\theta}{\cos\theta}$ is a trigonometric **identity** and is therefore always true.

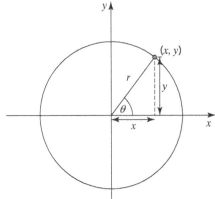

- Note from the diagram above that $\sin\theta = \dfrac{y}{r}$, $\cos\theta = \dfrac{x}{r}$, $\tan\theta = \dfrac{y}{x}$

$\tan\theta = \dfrac{y}{r} \times \dfrac{r}{x} = \sin\theta \times \dfrac{1}{\cos\theta} = \dfrac{\sin\theta}{\cos\theta}$, so $\tan\theta = \dfrac{\sin\theta}{\cos\theta}$

$\sin^2\theta + \cos^2\theta = 1$

- $\sin^2\theta + \cos^2\theta = 1$ is another trigonometric identity.
- Note from the diagram above that the equation of the circle is $x^2 + y^2 = r^2$.
- $x = r\cos\theta$, $y = r\sin\theta$ so $r^2\cos^2\theta + r^2\sin^2\theta = r^2$, so $\sin^2\theta + \cos^2\theta = 1$.

1) **Simplify** $\sin^2 4\alpha + \cos^2 4\alpha$.

   $\sin^2 4\alpha + \cos^2 4\alpha = 1$ ⟵ Note: This is the same as $\sin^2\theta + \cos^2\theta$ where θ has been replaced with 4α.

2) **Simplify** $3\sin^2\theta + 3\cos^2\theta$.

   $3\sin^2\theta + 3\cos^2\theta = 3(\sin^2\theta + \cos^2\theta)$

   $\qquad\qquad\qquad = 3 \times 1$

   $\qquad\qquad\qquad = 3$

3) **Simplify** $6 - 6\cos^2\theta$.

   $6 - 6\cos^2\theta = 6(1 - \cos^2\theta)$

   $\qquad\qquad\quad = 6 \times \sin^2\theta$

   $\qquad\qquad\quad = 6\sin^2\theta$

Key Point

$\tan\theta \equiv \dfrac{\sin\theta}{\cos\theta}$ for all values of θ (except where $\cos\theta = 0$).

Key Point

$\tan\theta \equiv \dfrac{\sin\theta}{\cos\theta}$ is not given in the formula booklet and must be learnt.

Key Point

$\sin^2\theta + \cos^2\theta = 1$ is not given in the formula booklet and must be learnt.

Key Point

$\sin^2\theta + \cos^2\theta = 1$ is sometimes known as Pythagoras' theorem in trigonometry.

Key Point

Always look for factors.

4) Simplify $\dfrac{\sin 2\theta}{\sqrt{1-\sin^2 2\theta}}$

$$\frac{\sin 2\theta}{\sqrt{1-\sin^2 2\theta}} = \frac{\sin 2\theta}{\sqrt{\cos^2 2\theta}}$$

Note: $\cos^2\theta = 1 - \sin^2\theta$ so $\cos^2 2\theta = 1 - \sin^2 2\theta$.

$$= \frac{\sin 2\theta}{\cos 2\theta}$$

$$= \tan 2\theta$$

5) Given that $\cos\theta = \dfrac{4}{5}$ and that θ is reflex, find the value of $\sin\theta$.

$$\sin^2\theta = 1 - \cos^2\theta$$

$$\sin^2\theta = 1 - \left(\frac{4}{5}\right)^2$$

$$\sin^2\theta = \frac{9}{25}$$

$$\sin\theta = \pm\frac{3}{5}$$

As θ is reflex and $\cos\theta$ is positive, θ must be in quadrant 4 and therefore $\sin\theta$ is negative, so $\sin\theta = -\dfrac{3}{5}$

> **Key Point**
>
> $\sin^2\theta + \cos^2\theta = 1$ can be written as:
>
> $\sin^2\theta = 1 - \cos^2\theta$
>
> or
>
> $\cos^2\theta = 1 - \sin^2\theta$

6) Show that $(\sin\theta + \cos\theta)^2 \equiv 1 + 2\sin\theta\cos\theta$.

$$\text{LHS} = (\sin\theta + \cos\theta)^2$$

$$= \sin^2\theta + 2\sin\theta\cos\theta + \cos^2\theta$$

$$= \sin^2\theta + \cos^2\theta + 2\sin\theta\cos\theta$$

$$= 1 + 2\sin\theta\cos\theta$$

$$= \text{RHS}$$

When completing a 'show that' question, always start with the left-hand side (LHS) or right-hand side (RHS) of the identity and work towards the other side.

Always look for $\tan\theta \equiv \dfrac{\sin\theta}{\cos\theta}$ or $\sin^2\theta + \cos^2\theta = 1$.

> **Quick Test**
>
> 1. Simplify $\dfrac{\sin\theta}{\cos\theta}$
>
> 2. Simplify $\dfrac{1}{2} - \dfrac{1}{2}\sin^2\theta$.
>
> 3. Given that $\tan\theta = \dfrac{5}{12}$ and that θ is obtuse, find the value of $\sin\theta$.
>
> 4. Show that $\tan x + \dfrac{1}{\tan x} \equiv \dfrac{1}{\sin x \cos x}$

> **Key Words**
>
> identity
> simplify
> show that

Solving Trigonometric Equations

You must be able to:

- Solve simple trigonometric equations in a given interval
- Solve quadratic trigonometric equations in sin, cos and tan in a given interval
- Solve trigonometric equations in sin, cos and tan involving multiples of the unknown angle.

Simple Trigonometric Equations in the Form $\sin x = k$, $\cos x = k$ and $\tan x = k$

- The first solution, α, to $\sin x = k$ is found using $\sin^{-1} k$ on your calculator.
- A second solution can be found from the symmetry of the sin graph: $180 - \alpha$ (in degrees as below) or $\pi - \alpha$ (in radians).
- Other solutions are then found by adding or subtracting multiples of $360°$ (or 2π radians).
- This approach can be extended to equations in the form $\cos x = k$ and $\tan x = k$.

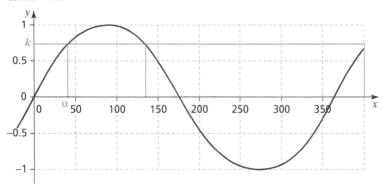

Key Point

When solving equations in the form $\sin x = k$, $\cos x = k$ or $\tan x = k$, your second solution can be found using the symmetry of the graph or using these rules:

	2nd solution
$\sin x$	180 or π – 1st solution
$\cos x$	360 or 2π – 1st solution
$\tan x$	180 or π + 1st solution

Solve the equation $\sin \alpha = \frac{\sqrt{3}}{2}$ in the **interval** $0° \leqslant \alpha \leqslant 360°$.

$\sin^{-1} \frac{\sqrt{3}}{2} = 60°$

$180 - 60 = 120°$ ◄

There are no other solutions as 60 ± 360 or 120 ± 360 will be outside the range $0° \leqslant \alpha \leqslant 360°$.

$\therefore \alpha = 60°$ and $120°$

Solving using a CAST diagram

$\sin^{-1} \frac{\sqrt{3}}{2} = 60°$ ◄

Therefore $\alpha = 60°$ and $120°$

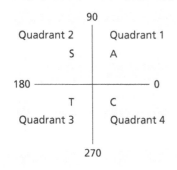

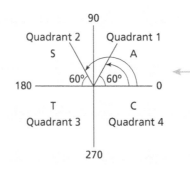

Key Point

When using a CAST diagram, always mark the angle with the x-axis.

Note the second solution is found by calculating 180 – first solution.

Note the first step is the same.

Mark the first solution on the CAST diagram. To find the second solution, find the other quadrant in which $\sin \alpha$ is positive (quadrant 2) and mark 60° with the x-axis in this quadrant.

To use the diagram, start at 0° and walk anti-clockwise (as you are looking for positive solutions) and write down solutions, i.e. 60° and 120°.

Solving Quadratic Trigonometric Equations

- When solving quadratic trigonometric equations, you may have to use a trigonometric identity to rearrange the equation before it can be solved.

Solve the equation $6\sin^2 x + \cos x = 4$, $0° \leqslant x \leqslant 360°$.

$6(1 - \cos^2 x) + \cos x = 4$

$6 - 6\cos^2 x + \cos x = 4$

$6\cos^2 x - \cos x - 2 = 0$

This is a quadratic equation of the form $6Y^2 - Y - 2 = 0$, which factorises to give:

$(3\cos x - 2)(2\cos x + 1) = 0$

So $\cos x = \frac{2}{3}$ and $\cos x = -\frac{1}{2}$

$x = \cos^{-1}\frac{2}{3} = 48.2°$ (3 s.f.)

Using the CAST diagram, $\cos x$ is positive in quadrant 4, so the second solution is $360 - 48.2 = 311.8°$.

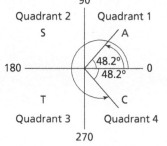

$\cos x = -\frac{1}{2}$

$x = \cos^{-1}\left(-\frac{1}{2}\right) = 120°$

Using the CAST diagram, $\cos x$ is negative in quadrant 3, so the second solution is $360 - 120 = 240°$.

Therefore $x = 48.2°$, $120°$, $240°$ and $311.8°$.

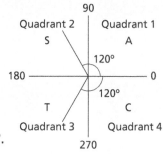

> **Key Point**
>
> In the majority of cases, you should rearrange the equation to be in terms of just one trigonometric function.

Use the trigonometric identity to write the equation in terms of $\cos x$.

This is a quadratic equation so make it $= 0$ before solving.

Note that $\cos x$ is negative in this case.

> **Key Point**
>
> When travelling anti-clockwise around a CAST diagram, you are finding positive solutions. When travelling clockwise, you are finding negative solutions.

Solving Quadratic Trigonometric Equations Involving Multiples of the Unknown Angle

- When solving equations in the form $\sin(n\theta + \alpha)$, $\cos(n\theta + \alpha)$ or $\tan(n\theta + \alpha)$, make sure you have all the solutions in the given interval.

Solve the equation $\tan\left(2\theta + \frac{\pi}{3}\right) = -0.3$, $-\pi \leqslant \theta \leqslant \pi$.

$-\pi \leqslant \theta \leqslant \pi$, $\therefore \frac{-5\pi}{3} \leqslant 2\theta + \frac{\pi}{3} \leqslant \frac{7\pi}{3}$

$2\theta + \frac{\pi}{3} = \tan^{-1}(-0.3) = -0.291$

$2\theta + \frac{\pi}{3} = -0.291 \Rightarrow \theta = -0.669$

$2\theta + \frac{\pi}{3} = -0.291 - \pi = -3.43 \Rightarrow \theta = -2.24$

$2\theta + \frac{\pi}{3} = -0.291 + \pi = 2.851 \Rightarrow \theta = 0.902$

$2\theta + \frac{\pi}{3} = -0.291 + 2\pi = 5.99 \Rightarrow \theta = 2.47$

$2\theta + \frac{\pi}{3} = -0.291 + 3\pi = 9.13 \Rightarrow \theta = 4.04$ (out of required range)

$\theta = -2.24$; -0.669; 0.902; 2.47

Note that θ is in radians.

You can adjust the range first.

Make the same changes to each side of the inequality as made to the centre.

The CAST diagram or the graph of $y = \tan x$ can now be used to find all solutions of $2\theta + \frac{\pi}{3}$

Quick Test

1. Solve the equation $\sin x = \frac{1}{2}$, $360° < x \leqslant 540°$.
2. Solve the equation $2\sin^2\theta + 5\cos\theta + 1 = 0$, $0 \leqslant \theta \leqslant 2\pi$.
3. Solve the equation $3\cos\left(\frac{1}{2}x + 45°\right) = 1$, $0° \leqslant x \leqslant 360°$.

> **Key Words**
>
> interval

Equations of Straight Lines

1 A(–2, –1) lies on line L_1 and line L_2.
   B(2, –7) and C(7, 5) lie on lines L_1 and L_2 respectively.

   a) Show that lines L_1 and L_2 are perpendicular.   [3]

   b) Find the equations of L_1 and L_2.   [2]

2 Show that the points A(–5, –3), B(–2, 2) and C(1, 7) are collinear.   [2]

3 Show that shape ABCD is a rectangle and find its area.   [6]

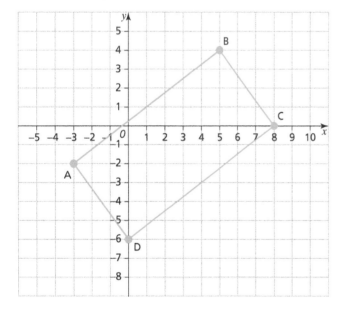

Circles

1 Show that the centre of a circle with equation $x^2 + y^2 + 2fx + 2gy + c = 0$ is $(-f, -g)$ and the radius is $\sqrt{f^2 + g^2 - c}$. [3]

2 A(-9, -3) and B(1, 7) lie on a circle with centre C.

Line AB forms a chord of the circle.

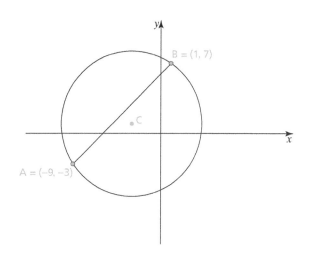

a) Show that the centre, C, lies on a line with the equation $y = -x - 2$. [3]

b) Given that the x-coordinate of C is -3, find the equation of the circle. [3]

Total Marks _____ / 9

Sine, Cosine & Tangent Functions, and Sine & Cosine Rule

1 Express cos230° as a trigonometric ratio of an acute angle. [2]

2 The diagram shows part of the graph $y = f(x)$. The graph crosses the x-axis at P(100, 0) and Q(a, 0).
The graph crosses the y-axis at R(0, b) and has a maximum value at S, as shown.

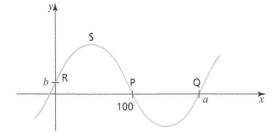

Given that $y = \sin(\theta + k)$, $k > 0$

a) Write down the value of a. [1]

b) Write down the smallest possible value of k. [1]

c) Write down the value of b. [1]

d) Write down the coordinates of S. [2]

3 The diagram shows a quadrilateral, ABCD.

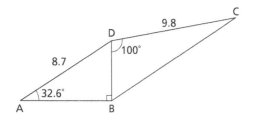

a) Find the length BC. [3]

b) Find the total area of ABCD. [4]

Total Marks .... / 14

Trigonometric Identities

1 Given that angle A is obtuse and $\cos A = -\sqrt{\frac{7}{11}}$, show that $\tan A = -\frac{2\sqrt{7}}{7}$ [3]

2 Simplify $\cos^4 x - \sin^4 x$. [3]

3 Show that $(1 + \sin\theta)^2 + \cos^2\theta \equiv 2(1 + \sin\theta)$. [4]

Total Marks .... / 10

Solving Trigonometric Equations

1 Solve $2\cos^2\theta + \sin\theta = 1$, $0° \leqslant \theta \leqslant 360°$. [4]

2 Solve $\sin 2\theta = \cos 2\theta$, $-360° \leqslant \theta \leqslant 0°$. [4]

3 Solve $3\tan\theta = 2\tan^2\theta$, $0° \leqslant \theta \leqslant 360°$. [4]

4 Solve $2\sin^2\left(\theta + \frac{\pi}{3}\right) = 1$, $0 \leqslant \theta \leqslant 2\pi$. [5]

5 **a)** Show that $\dfrac{\cos^2\theta}{\sin\theta + \sin^2\theta} = \dfrac{1 - \sin\theta}{\sin\theta}$ [3]

b) Hence solve the equation $\dfrac{\cos^2\theta}{\sin\theta + \sin^2\theta} = 2$, $0° \leqslant \theta \leqslant 180°$. [3]

Total Marks .... / 23

Review Questions

Indices and Surds

1 Simplify the following:

a) $\left(2a^4b^{\frac{3}{2}}\right)^{-2}$ [2]

b) $(3a^2b^3) \times 4a^3 \div ab^5$ [1]

c) $3^{\frac{1}{2}} + 3^{\frac{5}{2}} - 3^{\frac{3}{2}}$ [2]

2 Evaluate the following:

a) $(12^{-2})^{\frac{1}{2}}$ [1]

b) $\sqrt[3]{5^6 \times 15^{-9}}$ [2]

3 Rationalise the denominators of the following:

a) $\dfrac{2}{3-\sqrt{5}}$ [2]

b) $\dfrac{\sqrt{7}}{4+\sqrt{7}}$ [2]

> **Total Marks** .... / 12

Manipulating Algebraic Expressions

1 **a)** Write a simplified expression for the shaded area of the rectangle. [2]

b) Hence, write a factorised expression for the shaded area of the rectangle. [1]

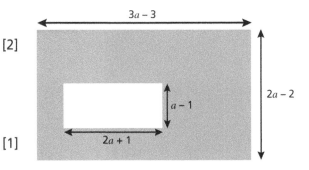

2 Factorise completely $3x - 12x^3$. [2]

3 Factorise completely $2x^2 - 4x - 6$. [2]

> **Total Marks** .... / 7

Expanding and Dividing Polynomials

1 Show that $x^4 + 2x^3 - 13x^2 - 14x + 24$ can be written in the form $(x + 4)(ax^3 + bx^2 + cx + d)$ and hence find the values of a, b, c and d. [4]

2 **a)** Write down the first four terms of the binomial expansion of $(2 - x)^{10}$ in ascending powers of x. [3]

b) Using the first four terms of the binomial expansion of $(2 - x)^{10}$, find an approximate value of 1.97^{10}. Give your answer to 4 decimal places. [2]

Total Marks .... / 9

Quadratic Equations

1 Show that the solutions to any quadratic equation $ax^2 + bx + c = 0$ are $x = \dfrac{-b \pm \sqrt{b^2 - 4ac}}{2a}$ [6]

2 Find the real roots of the function $y = x^8 - 18x^4 + 32$. [4]

3 Show that the graph of the function $f(x) = -2x^2 + 4x - 5$ does not intersect the x-axis. [2]

Total Marks .... / 12

Review Questions

Simultaneous Equations

1 Solve the simultaneous equations $2x + 3y = 8$ and $3x = 2y + 4$. [4]

2 Solve the simultaneous equations $y = 2x + 1$ and $(x + 1)^2 + (y - 1)^2 = 4$. [4]

3 **a)** Sketch a graph of $f(x) = x(x + 1)(x - 1)(x - 3)$ and $g(x) = -2x$. [3]

b) Show how you can use your sketch to find the number of real solutions to the equation $x^4 - 3x^3 - x^2 + 5x = 0$. [3]

Total Marks .... / 14

Inequalities

1 The graph shows $y = f(x)$ and $y = g(x)$ where $f(x) = 3x^2 - 5x - 2$ and $g(x) = 2x + 4$.

Find the values of x for which $f(x) > g(x)$. Express your answer in set notation. [3]

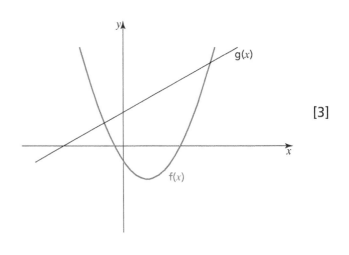

2 Draw a graph to show the region satisfied by the inequalities: [3]

$y > -\frac{3}{4}x + 4$

$y > \frac{2}{3}x + 1$

$y \leqslant 5$

Total Marks _____ / 6

Sketching Curves

1 Write down the equation shown by the graph in expanded and simplified form and estimate the solution(s). [6]

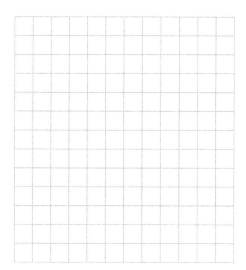

$y = (x - 2)(x - 4)$

$y = -x(x + 1)(x - 1)(x - 3)$

2 The table shows the results of an experiment of heating water over a period of time.

Time (min)	0	3	6	9	12	15
Temperature (°C)	0	7	25	65	98	100

Explain whether or not the relationship between time and temperature is direct proportion. [2]

Total Marks _____ / 8

Translating Graphs

1 The point A(–2, 8) lies on the curve $y = f(x)$. The coordinates of A after a translation are (3, 6).

Write down the transformation in the form $y = f(x + a) + b$. [3]

2 Given $f(x) = x^2$, write down the value of the turning point and sketch the graph of $y = f(x – 2) – 3$. [2]

3 Given the graph of $f(x) = x(x + 1)(x – 3)$ and $y = f(x + a)$ below, write down the value of a and hence the equation of $y = f(x + a)$. [2]

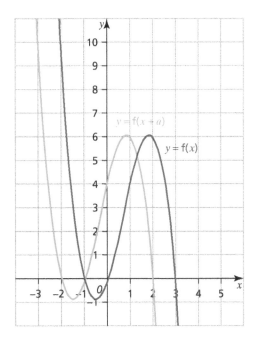

Total Marks .... / 7

Stretching and Reflecting Graphs

1 The point A(–2, 5) lies on the curve $y = f(x)$. Write down the coordinates of point A after the transformation:

a) $y = f\left(\frac{1}{2}x\right)$ [1]

b) $4y = f(x)$ [2]

2 Given the graph of $y = f(x)$:

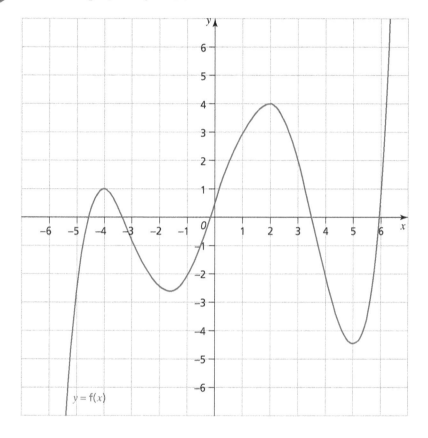

a) Sketch the transformation of $y = -f(x)$ on the grid above. [2]

b) Sketch the transformation of $y = f(-x)$ on the grid above. [2]

Exponentials and Logarithms

You must be able to:

- Use the functions a^x $(a > 0)$ and $\log_a x$ and sketch their graphs
- Use the functions e^x and $\ln x$ and sketch their graphs
- Know that the gradient of e^{kx} is equal to ke^{kx}
- Solve simple exponential and logarithmic equations.

Exponential Functions and their Inverses

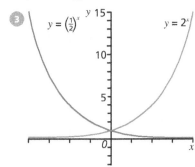

Exponential Functions $f(x) = a^x$

- The function $f(x) = a^x$ $(a > 0)$ has domain $x \in \mathbb{R}$ and range $f(x) \in \mathbb{R}, f(x) > 0$.
- $f(x) = a^x$ $(a > 0)$ has an **asymptote** at $y = 0$.
- The graph of $y = a^x$ $(a > 0)$ always intersects the y-axis at the point $(0, 1)$. See graph ❶.
- The **inverse function** of $f(x) = a^x$ is $f^{-1}(x) = \log_a x$ (i.e. the logarithm of x to base a).
- As with all inverse functions, this can be found by reflecting the original function in the line $y = x$ (see graph ❷).
- The domain of $f^{-1}(x) = \log_a x$ is $x \in \mathbb{R}, x > 0$ and the range is $f^{-1}(x) \in \mathbb{R}$.
- $f^{-1}(x) = \log_a x$ has an asymptote at $x = 0$.
- The graph of $y = \log_a x$ always intersects the x-axis at the point $(1, 0)$.
- When $a < 0$, the graph of $y = a^x$ has a different shape. For example, compare $y = 2^x$ and $y = \left(\dfrac{1}{2}\right)^x$. See graph ❸.
- Notice that $y = \left(\dfrac{1}{2}\right)^x = \dfrac{1}{2^x} = 2^{-x}$. So $f(x) = 2^{-x}$ is a reflection of $f(x) = 2^x$ in the y-axis.

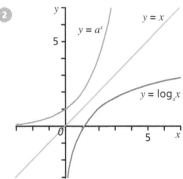

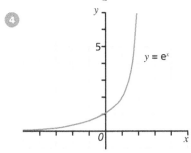

'The' Exponential Function $f(x) = e^x$

- $f(x) = e^x$ is often referred to as '**the**' exponential function.
- e is an irrational number, and e is approximately equal to 2.71828…
- The graph of $y = e^x$ looks like all other exponential functions, intersecting the y-axis at the point $(0, 1)$. See graph ❹.
- What is special about $y = e^x$ is that it is equal to its own derivative. In other words, if $y = e^x$, then $\frac{dy}{dx} = e^x$. You are not expected to prove this.
- For example, the gradient at the point on the graph of $y = e^x$ where $x = 1.5$ is equal to $e^{1.5} \approx 4.5$ (see graph ❺).
- The inverse function of $f(x) = e^x$ is $f^{-1}(x) = \log_e x$. This is commonly written as $f^{-1}(x) = \ln x$.
- The graph of $y = \ln x$ may be illustrated by reflecting $y = e^x$ in the line $y = x$.

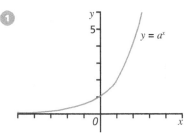

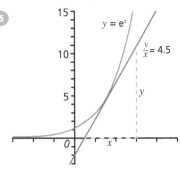

- The domain of $f^{-1}(x) = \ln x$ is $x \in \mathbb{R}$, $x > 0$ and the range is $f^{-1}(x) \in \mathbb{R}$.
- $f^{-1}(x) = \ln x$ has an asymptote at $x = 0$.
- The graph of $y = \ln x$ intersects the x-axis at the point $(1, 0)$. See graph ❻.

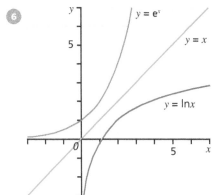

The Function $f(x) = e^{kx}$

- $f(x) = e^{kx}$ is a transformation of $f(x) = e^x$; namely a stretch of scale factor $\frac{1}{k}$ parallel to the x-axis. For example, see graph ⑦.
- What is special about $y = e^{kx}$ is that its derivative is ke^{kx}, or ky. In other words, if $y = e^{kx}$, then $\frac{dy}{dx} = ke^{kx}$. You are not expected to prove this.
- For example, to find the gradient at the point where $x = 0.2$ on the curve $y = e^{2x}$:
  $\frac{dy}{dx} = 2e^{2x}$, so at $x = 0.2$ the gradient will be equal to $2e^{0.4} \approx 2.98$.
- Since $\frac{dy}{dx} = ky$, you can also write this as $\frac{dy}{dx} \propto y$. This means that the rate of change is proportional to the y-value.

⑦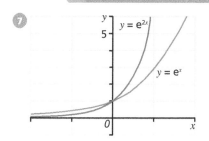

Exponential Models

- Whenever $\frac{dy}{dx} \propto y$ occurs in a real-life situation, it is appropriate to use an **exponential model** for the data.
- So working backwards, if $\frac{dy}{dx} \propto y$, then a suitable mathematical model for the data would be to assume $y = Ae^{kx}$, where A and k are constants.
- For example, during the observation of a population of rabbits, R, suppose the rate of increase of the rabbit population $\frac{dR}{dt}$ was found to be proportional to the number of rabbits at any particular time. In other words, that $\frac{dR}{dt} \propto R$.
- Therefore, in cases such as this, the relationship between the number of rabbits, R, and the time, t, will be $R = Ae^{kt}$, where A and k are constants. The constants would depend on the initial population of rabbits and how quickly they are breeding.

Solving Equations

Solving Equations of the Type $e^{ax+b} = p$

Solve the equation $e^{2x-3} = 5$.

$2x - 3 = \ln 5$ ← Use the fact that the inverse function of e^x is $\ln x$.

$2x = 3 + \ln 5$

$x = \frac{3 + \ln 5}{2}$ ← Now just rearrange to find x.

Solving Equations of the Type $\ln(ax + b) = q$

Solve the equation $\ln(4 - 2x) = 1$.

$4 - 2x = e^1$ ← Use the fact that the inverse function of $\ln x$ is e^x.

$2x = 4 - e$

$x = \frac{4 - e}{2}$ ← Now rearrange to find x.

> **Key Point**
>
> Always check when solving equations such as these that you don't end up with the logarithm of a negative number. For example, $e^{2x-3} = -5$ would lead to $2x - 3 = \ln(-5)$, so no solution would exist.

Quick Test

1. Sketch the graph of $y = \ln(x - 2)$, stating any points of intersection with the x- and y-axes.
2. Solve the equation $3e^{4x-1} = 2$.
3. Solve the equation $\ln(3x + 2) = -1$.
4. Find the gradient on the curve $y = e^{6x}$ at the point $\left(\frac{1}{3}, e^2\right)$.

> **Key Words**
>
> asymptote
> inverse function
> exponential function
> exponential model

Laws of Logarithms

You must be able to:

- Understand and use the laws of logarithms
- Solve exponential equations of the form $a^x = b$.

Laws of Logarithms

- There are six main logarithm rules that are true for all **bases**. In each of the statements, a is the base of the logarithm.

Rule 1	$\log_a x + \log_a y = \log_a(xy)$
Rule 2	$\log_a x - \log_a y = \log_a\left(\dfrac{x}{y}\right)$
Rule 3	$\log_a x^k = k\log_a x$
Rule 4	$\log_a b = \dfrac{\log_c b}{\log_c a}$
Rule 5	$\log_a a = 1$
Rule 6	$\log_a 1 = 0$

The 'change of base' formula.

Evaluating Expressions Involving Logarithms

- You can use the rules of logarithms to evaluate expressions without the use of a calculator.

Work out the following, without using a calculator:

a) $\log_2 16 + \log_4 64$

You can write 16 = 2^4 and 64 = 4^4 in anticipating using rule 5.

$\log_2 16 + \log_4 64$
$= \log_2 2^4 + \log_4 4^4$
$= 4\log_2 2 + 4\log_4 4$

Now rule 5 can be used.

$= 4 + 4$
$= 8$

b) $\log_6 72 - \log_6 2$

$\log_6 72 - \log_6 2$
$= \log_6\left(\dfrac{72}{2}\right)$

Use rule 2 to combine the logarithms.

$= \log_6 36$
$= \log_6 6^2$
$= 2\log_6 6$

Use rule 3.

$= 2$

c) $\log_5 4 - 2\log_5 3 + \log_5\left(\dfrac{9}{4}\right)$

$\log_5 4 - 2\log_5 3 + \log_5\left(\dfrac{9}{4}\right)$
$= \log_5 4 - \log_5 3^2 + \log_5\left(\dfrac{9}{4}\right)$
$= \log_5 4 - \log_5 9 + \log_5\left(\dfrac{9}{4}\right)$
$= \log_5\left(\dfrac{4}{9}\right) + \log_5\left(\dfrac{9}{4}\right)$

Use rule 1 to combine the fractions.

$= \log_5\left(\dfrac{36}{36}\right)$
$= \log_5 1$

Easy finish by using rule 6.

$= 0$

Key Point

It is important to remember that $x = a^y \Leftrightarrow y = \log_a x$. In other words, the statements $x = a^y$ and $y = \log_a x$ are interchangeable. They say exactly the same thing.

Key Point

Logarithms to base 10 are called **common logarithms**, and written as $\log_{10}$ or simply lg.

Logarithms to base e are called **natural logarithms** and written as $\log_e$ or simply ln.

Solving Equations Involving Logarithms

- Again, the technique is to simplify the equation using the rules of logarithms in a sequence of logical steps. The following examples will show how most of the rules may be used.

Solve the following equations:

a) $\log_{16} x - \log_{16} 10 = \frac{1}{2}$

$$\log_{16}\left(\frac{x}{10}\right) = \frac{1}{2}$$

$$\frac{x}{10} = 16^{\frac{1}{2}}$$

$$\frac{x}{10} = 4 \text{, so } x = 40$$

b) $\log_3(x-1) - \log_3 2 = 2 + \frac{1}{\log_9 3}$

$$\log_3(x-1) - \log_3 2 = 2 + \frac{1}{\log_9 3}$$

$$\log_3(x-1) - \log_3 2 = 2 + \log_3 9$$

$$\log_3\left(\frac{x-1}{2}\right) = 2\log_3 3 + \log_3 9$$

$$\log_3\left(\frac{x-1}{2}\right) = \log_3 3^2 + \log_3 9$$

$$\log_3\left(\frac{x-1}{2}\right) = \log_3 9 + \log_3 9$$

$$\log_3\left(\frac{x-1}{2}\right) = \log_3 81$$

$$\frac{x-1}{2} = 81$$

$$x - 1 = 162 \text{, so } x = 163$$

c) $\log_3 2 + 2\log_3(x-2) = \log_3 48$

$$\log_3 2 + 2\log_3(x-2) = \log_3 48$$

$$\log_3 2 + \log_3(x-2)^2 = \log_3 48$$

$$\log_3[2(x-2)^2] = \log_3 48$$

$$2(x-2)^2 = 48$$

$$(x-2)^2 = 24$$

$$x - 2 = \pm\sqrt{24}$$

$$x = 2 \pm \sqrt{24}$$

$$\Rightarrow \text{ only solution is } x = 2 + \sqrt{24}$$

Key Point

Example **b)** makes use of the fact that $\log_c b = \frac{1}{\log_b c}$. This is often very useful to remember.

It is important to note that $x = 2 - \sqrt{24}$ is a negative number, therefore it is not a possible solution to this equation, since $2\log_3(x-2)$ is undefined at this value of x.

Use rule 4 with $b = c$.

Since $\log_3 3 = 1$.

Now we use the fact that if $\log_a x = \log_a y$, then $x = y$.

Exponential Equations of the Form $a^x = b$

- To solve these types of equation, you 'take logarithms of both sides'.

1) Solve the equation $10^x = 5$.

So $\log_{10} 10^x = \log_{10} 5$

$x \log_{10} 10 = \log_{10} 5$

$x = \log_{10} 5$

$x = 0.699$

In this case, since 10 is the base, it makes sense to take logarithms to base 10.

2) Solve the equation $4^{2x+1} = 5$.

$\log_4 4^{2x+1} = \log_4 5$

$(2x+1)\log_4 4 = \log_4 5$

$2x + 1 = \log_4 5$

So $2x = -1 + \log_4 5$

And $x = \frac{-1 + \log_4 5}{2} = 0.0805$

Taking logarithms to base 4.

- If your calculator can only calculate logarithms using bases e or 10, use the 'change of base' formula.

- So in the example above right, $x = \dfrac{-1 + \dfrac{\log_{10} 5}{\log_{10} 4}}{2} = 0.0805$

Quick Test

1. Evaluate $\log_2 16 - \log_2\left(\frac{1}{2}\right) + \log_8 64$.
2. Solve the equation $2^{5-x} = 50$.
3. Solve the equation $2(\lg x)^2 - 7\lg x + 3 = 0$.

Key Words

base
change of base formula
common logarithm
natural logarithm

Modelling Using Exponentials and Logarithms

You must be able to:

- Use logarithmic graphs for relationships such as $y = ax^n$ and $y = kb^x$
- Understand and use exponential growth and decay.

Logarithmic Graphs

Using Logarithmic Graphs for Relationships of the Form $y = ax^n$

- Given a **relationship** between two variables x and y of the form $y = ax^n$, you can estimate the values of the **constants** a and n by plotting $\log_{10}x$ on the x-axis and $\log_{10}y$ on the y-axis.
- Start with the relationship: $y = ax^n$

$$\log_{10} y = \log_{10}(ax^n)$$
$$\log_{10} y = \log_{10}a + \log_{10}x^n$$
$$\log_{10} y = \log_{10}a + n\log_{10}x$$
$$Y = c + mX$$

Key Point

You could actually take logarithms to any base, though base 10 is most commonly used in this kind of relationship.

Now take logarithms of both sides (to base 10).

Now compare this with the equation of a straight line $y = mx + c$.

- So if you plot $\log_{10}x$ on the x-axis, and plot $\log_{10}y$ on the y-axis, the gradient of the graph will be an estimate for n and the y-intercept will give $\log_{10}a$.

It is known that when an object falls from rest, through a height h, the time taken t is related to h by the formula $h = kt^n$.

In an experiment carried out by a student, the results in the table (right) were obtained after an object was dropped from rest, from various heights h. By using the student's results, and drawing a suitable straight-line graph, determine approximate values for k and n.

t (seconds)	0.65	0.90	1.09	1.31	1.44
h (metres)	2	4	6	8	10

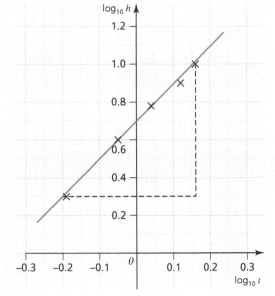

First calculate the required table of values, using the data given. Plot these values on a graph.

You need to plot $\log_{10}t$ on the x-axis and $\log_{10}h$ on the y-axis.

$\log_{10} t$	−0.19	−0.05	0.04	0.12	0.16
$\log_{10} h$	0.30	0.60	0.78	0.90	1.00

From the graph, the gradient

$= \dfrac{0.7}{0.36} = 1.94$

This suggests that $n = 2$.

From the graph, the y-axis intercept is $c = 0.7$.

So $0.7 = \log_{10}a$.

Therefore $a = 10^{0.7} = 5.01$, suggesting a is approximately equal to 5.

So, from the student's results, the suggested relationship is $h = 5t^2$.

Using Logarithmic Graphs for Relationships of the Form $y = kb^x$

- Given a relationship $y = kb^x$, you can estimate the values of the constants k and b by plotting x on the x-axis and $\log_{10}y$ on the y-axis.

$$y = kb^x$$
$$\log_{10} y = \log_{10}(kb^x) \longleftarrow$$
$$\log_{10} y = \log_{10}k + \log_{10}b^x$$
$$\log_{10} y = \log_{10}k + x\log_{10}b \longleftarrow$$
$$Y = c + Xm$$

Now take logarithms of both sides (to base 10).

Again compare this with the equation of a straight line $y = mx + c$.

- So if you plot x on the x-axis, and plot $\log_{10}y$ on the y-axis, the gradient of the graph will be an estimate for $\log_{10}b$ and the y-intercept will give $\log_{10}k$.

A new smartphone is introduced to the market. After several years, the sales figures ('y' in millions) are compared against the number of years (t) since the release date. The results are given in the table.

Year(s) (t) after release date	1	2	3	4	5
Number of sales (y) in millions	3.8	8.8	18.6	35.2	69.1

a) It is suggested that y and t are connected by the relationship $y = kb^t$ where k and b are constants. By using the sales figures, and drawing a suitable straight-line graph, determine the values of k and b.

First calculate the required table of values, using the data given. Plot these values on a graph.

This time, you need to plot t on the x-axis and $\log_{10}y$ on the y-axis.

Year (t)	1	2	3	4	5
$\log_{10}y$	0.58	0.94	1.27	1.55	1.84

From the graph, the gradient $= \dfrac{1.7 - 0.3}{5 - 0} = 0.28$.

So $0.28 = \log_{10}b$.

Therefore $b = 10^{0.28} = 1.91$.

From the graph, the y-axis intercept is $c = 0.3$.

So $0.3 = \log_{10}k$.

Therefore $k = 10^{0.3} = 2.00$.

So, from the student's results, the suggested relationship is $y = 2.00 \times 1.91^t$.

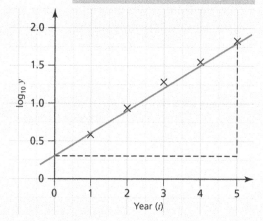

b) Suggest two reasons why this relationship is not likely to be the correct one.

This is not likely to be the correct relationship, since it indicates that sales of the smartphone will continue to increase indefinitely. Also, substituting $t = 0$ into the equation $y = 2.00 \times 1.91^t$ gives $y = 2.00$, suggesting there were 2 million sales at launch date, which is, of course, impossible.

Exponential Growth and Decay

- The example above illustrates exponential growth. Both this and exponential decay occur in many areas of mathematics and physics.
- In radioactive decay, it can be shown that the amount of radioactive substance N is related to time t by the equation $N = N_0 e^{-\lambda t}$.
- λ is known as the decay constant and N_0 is the initial amount of radioactive material (i.e. the amount of material at time $t = 0$).

Key Point

It is important to note that in questions such as this, models are just that – models. They may need to be further refined, and you may also be asked for suggestions as to why and how this should be done.

Quick Test

In each of these relationships, suggest what should be plotted on the x- and y-axes to obtain a straight line. In each case, a and b are constants.
1. $y = ae^{bx}$
2. $y = a\log_z x + b$
3. $y = ax + bx^3$
4. $y = a\log_z x + bx$

Key Words

relationship
constant
exponential growth
exponential decay

Differential of a Polynomial

You must be able to:

- Understand that $\frac{dy}{dx}$ is the rate of change of y with respect to x
- Differentiate kx^n for rational values of n
- Differentiate from first principles for small positive integer powers of x
- Understand and use the second derivative as the rate of change of gradient.

The Gradient of a Curve

- The gradient of a straight line $y = mx + c$ is the constant m.
  As x increases by 1, y increases by m.
- The gradient of a curve changes as you move along the curve.
- The gradient represents the rate of change of y with respect to x.

Differentiation from First Principles

- The gradient of a curve at a specific point is defined as being the same as the gradient of the tangent to the curve at that point.
- For example, the diagram shows a section of the curve $y = x^2$. Two points $P(x, x^2)$ and $Q(x + h, (x + h)^2)$ join to form the chord PQ:

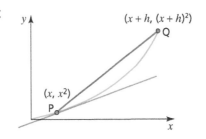

> **Key Point**
>
> If $y = f(x)$ then the gradient function is defined as $f'(x)$.
> $f'(x) = \frac{dy}{dx}$

- The gradient of PQ $= \frac{(x + h)^2 - x^2}{(x + h) - x}$

Using the gradient formula for a straight line.

$$= \frac{x^2 + 2hx + h^2 - x^2}{x + h - x}$$

Expand, simplify and factorise the numerator.

$$= \frac{h(2x + h)}{h}$$

Cancel the factor h.

$$= 2x + h$$

- As $h \to 0$ the gradient of the chord becomes closer to the gradient of the tangent and so to the gradient of the curve at that point.

As $h \to 0$ $2x + h \to 2x$

So $f(x) = x^2$ $f'(x) = 2x$

> **Key Point**
>
> $\frac{dy}{dx}$ = the change in y per unit increase in x.

Differentiation of kx^n

- If $y = x^n$ then $\frac{dy}{dx} = nx^{n-1}$, or using function notation if $f(x) = x^n$ then $f'(x) = nx^{n-1}$

1) Find $\frac{dy}{dx}$ when:

a) $y = 14x + 3$

Note this is the equation of a straight line. Remember $14x = 14x^1$

$\frac{dy}{dx} = 14$

$1 \times 14x^0 = 14$

b) $y = x^2 + 7x - 4$

You can differentiate each term separately.

$\frac{dy}{dx} = 2x + 7$

$2 \times x^1 = 2x$, $1 \times 7x^0 = 7$ and -4 differentiates to zero.

c) $y = \frac{1}{2}x^{-3} + 2x^{\frac{1}{2}}$

$\dfrac{dy}{dx} = -\dfrac{3}{2}x^{-4} + x^{-\frac{1}{2}}$

> Remember, when dealing with negative and fractional powers you still need to subtract 1.

d) $y = 3\sqrt{x} - 5x$

$y = 3x^{\frac{1}{2}} - 5x$

$\dfrac{dy}{dx} = \dfrac{3}{2}x^{-\frac{1}{2}} - 5$

> You need to express the function in power form before you can differentiate.

2) Given that the volume (V) of an expanding sphere is related to the radius (r) by the formula $V = \frac{4}{3}\pi r^3$, find the rate of change of the volume with respect to the radius at the instant when the radius is 6 cm.

$V = \frac{4}{3}\pi r^3$

$\dfrac{dV}{dr} = 4\pi r^2$

> The rate of change of V with respect to r is $\dfrac{dV}{dr}$

When $r = 6$, $\dfrac{dV}{dr} = 4 \times \pi \times 6^2 = 452$ (3 s.f.)

So the rate of change is 452 cm³ per cm.

3) Sketch the graph of the gradient function of $y = x^3$.

If $y = x^3$ $\dfrac{dy}{dx} = 3x^2$

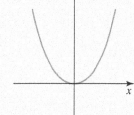

The Second Derivative: $\dfrac{d^2y}{dx^2}$ or $f''(x)$

- The second derivative is the rate of change of $\frac{dy}{dx}$ with respect to x.
- To find the second derivative, you need to differentiate twice.

Find $\dfrac{d^2y}{dx^2}$ when $y = 3x^7 + \dfrac{1}{x^4}$

$y = 3x^7 + x^{-4}$

$\dfrac{dy}{dx} = 21x^6 - 4x^{-5}$

$\dfrac{d^2y}{dx^2} = 126x^5 + 20x^{-6}$

> Remember, $- \times - = +$

> **Key Point**
>
> Always write your function in power form before starting to differentiate.

Quick Test

1. Find $\dfrac{dy}{dx}$ when $y = 5x^6 + 8x - 2$.
2. Find the value of $\dfrac{dy}{dx}$ when $y = \dfrac{3x + 8}{x^2}$.
3. Find $f''(x)$ when $f(x) = 3x^3(7x - 5x^3)$.
4. Find the points on the curve $y = 3x^3 - 5x^2$ where the gradient is 1.

> **Key Words**
>
> gradient
> rate of change
> tangent

Stationary Points, Tangents and Normals

You must be able to:

- Use differentiation to find equations of tangents and normals at specific points on a curve
- Identify where functions are increasing or decreasing
- Use differentiation to find maxima, minima and stationary points.

Increasing and Decreasing Functions

This is a graph of an **increasing function**. The gradient is always positive. 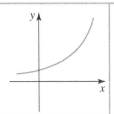	This is a graph of a **decreasing** function. The gradient is always negative. 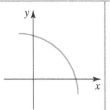	This graph is increasing for $x < 0$ and decreasing for $x > 0$. 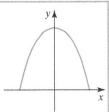

Find the values of x for which $f(x) = 3x^2 + 8x + 2$ is increasing.

$\dfrac{dy}{dx} > 0$ ← For a function to be increasing, the gradient should be positive.

$\dfrac{dy}{dx} = 6x + 8$

$6x + 8 > 0$, so $x > -\dfrac{4}{3}$

$3x^2 + 8x + 2$ is increasing for $x > -\dfrac{4}{3}$

Maximum, Minimum and Stationary Point

- Any point where the function stops increasing and starts decreasing is called a **maximum**.
- Any point where the function stops decreasing and starts increasing is called a **minimum**.
- These points are collectively called **stationary points** or **turning points**.
- At stationary points or turning points the gradient is zero.
- **Points of inflection** can also be stationary points.

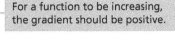

Key Point

If $\dfrac{dy}{dx} = 0$ and $\dfrac{d^2y}{dx^2} < 0$, maximum.

If $\dfrac{dy}{dx} = 0$ and $\dfrac{d^2y}{dx^2} > 0$, minimum.

If $\dfrac{dy}{dx} = 0$ and $\dfrac{d^2y}{dx^2} = 0$, but $\dfrac{d^3y}{dx^3} \neq 0$, point of inflection.

Key Point

The product of the gradients of perpendicular lines $= -1$

$m_1 \times m_2 = -1$

Find the nature of the stationary points of the curve with equation $y = x^4 + 4x^3 - 6$. Hence sketch the curve.

$y = x^4 + 4x^3 - 6$

$\dfrac{dy}{dx} = 4x^3 + 12x^2$ ←

$4x^3 + 12x^2 = 0$, $x^2(4x + 12) = 0$

$x = 0$, $y = -6$ **and** $x = -3$, $y = -33$ for the stationary points.

$\dfrac{d^2y}{dx^2} = 12x^2 + 24x$

At $x = 0$, $\dfrac{d^2y}{dx^2} = 0$ ∴ point of inflection. At $x = -3$, $\dfrac{d^2y}{dx^2} = 36 > 0$ ∴ minimum.

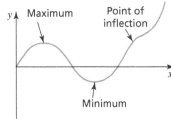

To find stationary points, set $\dfrac{dy}{dx} = 0$.

To sketch the curve, plot the minimum point and the point of inflection.

Also as $x \to \pm\infty$, $y \to \infty$.

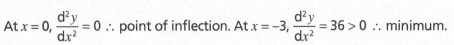

Equations of Tangents and Normals

- The equation of the **tangent** to the curve at the point where $x = a$ can be found using the formula $y - y_1 = m(x - x_1)$ (see page 30) with $m = \dfrac{dy}{dx}$.
- The **normal** to the curve at the point $x = a$ is the line which goes through a and is perpendicular to the tangent.
- The equation of the normal to the curve at the point where $x = a$ can be found using the formula $y - y_1 = m(x - x_1)$ with $m = \dfrac{-1}{dy/dx}$.

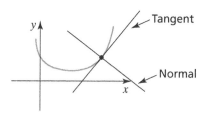

Find the equation of the tangent and the normal to the curve $y = x^3 - 3x^2 + 2x - 1$ at the point (3, 5).

$\dfrac{dy}{dx} = 3x^2 - 6x + 2$ ⟵ | The gradient of the curve is the same as the gradient of the tangent at any particular point.

At $x = 3$, $\dfrac{dy}{dx} = 3 \times 3^2 - 6 \times 3 + 2 = 11$

Equation of tangent is $y - 5 = 11(x - 3)$ ∴ $y = 11x - 28$

Gradient of tangent = 11; gradient of the normal = $-\dfrac{1}{11}$ ⟵ | To find the equation of the normal, find the perpendicular gradient.

Equation of the normal is $y - 5 = -\dfrac{1}{11}(x - 3)$ ∴ $11y + x - 58 = 0$

Problem Solving with Differentiation

Key Point

When dealing with differentiation problems of this nature, you will often need simultaneous equations.

A large tank in the shape of a cuboid is made from 60 cm² of sheet metal. The tank has no lid, and has length and width y cm and height x cm. A diagram of the tank is shown.

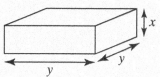

a) Show that the volume, V cm³, of the tank is given by $V = 15y - \dfrac{y^3}{2}$.

$V = y^2x$ and $SA = 4xy + 2y^2 = 60$ ⟵ | Write down two equations based on the information given.

$x = \dfrac{30 - y^2}{2y}$ ⟵ | Make x the subject of the second equation.

$V = y^2\left(\dfrac{30 - y^2}{2y}\right)$ ⟵ | Substitute into the first equation to eliminate x.

$V = \dfrac{30y^2}{2y} - \dfrac{y^4}{2y}$, so $V = 15y - \dfrac{y^3}{2}$ ∴ shown.

b) Use calculus to find the maximum value of V.

$\dfrac{dV}{dy} = 15 - \dfrac{3}{2}y^2 \qquad 15 - \dfrac{3}{2}y^2 = 0, \ y = \sqrt{10}$

$V = 15\sqrt{10} - \dfrac{\left(\sqrt{10}\right)^3}{2} = 31.6\,\text{cm}^3$ (3 s.f.)

Key Words

increasing function
decreasing function
maximum
minimum
point of inflection
stationary point
normal
calculus

Quick Test

1. Find the values for which $f(x) = x^2 - 9x$ is decreasing.
2. Find the coordinate of the stationary point on the curve with equation $y = 9 + x - x^2$ and determine the nature of the turning point.
3. Find the equation of the tangent and normal to the curve with equation $y = x^2 - 5x$ at the point (6, 6).

Exponentials and Logarithms

1 Sketch the graph of $y = 2^x - 2$, showing clearly any asymptotes and intersections with any axes. [4]

2 a) Find the gradient of the graph of $y = e^{kx}$ at the point on the graph where $x = \frac{1}{k}$ [3]

b) Hence show that the tangent line at this point has equation $y = kex$. [2]

3 Solve the equation $e^{\frac{x}{2}} = \frac{1}{2}$, giving your answer in the form $x = a\ln b$. [3]

Total Marks .... / 12

Laws of Logarithms

1 Solve the equation $3^{4x} = 45$, giving your answer to 3 significant figures. [3]

2 Simplify the expression $\log_3 4 - \log_3 36$, giving your answer as an integer. [5]

3 Solve the equations:

a) $(2^x)^2 - 2^x - 6 = 0$ [4]

b) $\log_{10} a + \log_a 10 = \frac{10}{3}$ [6]

Total Marks .... / 18

Modelling Using Exponentials and Logarithms

1 A population of lemurs, P, after t months is given by the formula $P = Ae^{0.2t}$.

Find the time taken for the population to double from its initial value. [4]

2 It is suggested that two variables, x and y, are related by the equation $y = A \times 10^{kx}$.

Given that $y = 45$ when $x = 2$, and $y = 45\,000$ when $x = 8$, determine the values of A and k. [4]

3 It is suggested that two variables, x and y, are related by the equation $y = ax^n$.

A student plots a graph of $\log_{10} y$ against $\log_{10} x$. By drawing a line of best fit, she estimates the gradient to be 1.48 and the y-intercept of the line to be at −0.60.

From these results, determine the likely values of a and n. [6]

Total Marks .... / 14

Practice Questions

Differential of a Polynomial

1 The function $f(x) = x + \frac{9}{x}, x \neq 0$.

    a) Find $f'(x)$.   [2]

    b) Solve $f'(x) = 0$.   [2]

2   **a)** Differentiate $y = 2x^3 + \sqrt{x} + \frac{x^2 + 2x}{x^2}$ with respect to x.   [3]

    b) Find the value of $\frac{dy}{dx}$ when $x = 5$.   [2]

3 Find $\frac{dy}{dx}$ and $\frac{d^2y}{dx^2}$ when $y = (2x + 4)(3x - 5)$.   [5]

4 Find the second derivative for the function $f(x) = 4x^3 - 12x^2 + 5$.   [2]

Total Marks .... / 16

Stationary Points, Tangents and Normals

1 A curve has equation $y = x^3 - 6x^2 + 9x$.

    a) Find the coordinates of the turning points.   [4]

    b) Determine the nature of the turning points.   [2]

2 Find the equation of the tangent to the curve with equation $y = x^2 + 4x - 3$ at the point (3, 18). [3]

3 Find the equation of the normal to the curve with equation $y = 2\left(1 - \frac{1}{x^2}\right)$ when $x = -4$. [4]

4 If $y = x^4 - 8x^3 - 62x^2 + 144x + 300$

   a) Find $\frac{dy}{dx}$ [2]

   b) Show that there is a stationary point at $x = 1$ and find the other two stationary points. [3]

   c) Sketch the shape of the curve. [2]

5 Find the values for which the function $f(x) = 5x^2 + 12x$ is increasing. [3]

Total Marks _____ / 23

Equations of Straight Lines

1 Find the equation of the line l that passes between the points (5, 9) and (11, 12).

Express your answer in the form $ax + by + c = 0$. [4]

2 The lines L_1, $y = \frac{2}{3}x - 4$, and L_2, $y - 7 = m(x - 10)$, where m is the gradient of L_2, are perpendicular and meet at point A.

Find the coordinates of point A. [5]

Total Marks .... / 9

Circles

1 Show that point A(–6, 0) lies on the circle $x^2 + y^2 + 6x - 2y + 10 = 10$ and find the equation of the tangent to the circle at that point. [4]

2 Find the equation of a circle with centre C(–2, 8) that passes through point A(–5, 12). [2]

3 The points A(–2, 5), B(6, 9) and C(3, 0) form the vertices of a triangle.

Find the equation of the circumcircle. [8]

Total Marks .... / 14

Sine, Cosine & Tangent Functions, and Sine & Cosine Rule

1   **a)** Sketch the graph of $y = 4\sin x$ in the range $0 \leqslant x \leqslant 2\pi$.   [2]

   b) Write down the coordinates of the maximum and minimum points.   [2]

2   The diagram shows part of a curve with equation $y = \cos(ax + b)$, where $a > 0$ and $0 < b < \pi$. The curve crosses the x-axis at the points Q, R and S.

Given that the coordinates of Q, R and S are $\left(\frac{\pi}{8},0\right),\left(\frac{5\pi}{8},0\right)$ and $\left(\frac{9\pi}{8},0\right)$ respectively, find the values of a and b.   [3]

3   The diagram shows triangle ABC with length AB = 30 cm, BC = 40 cm and area 400 cm².

Work out the value of angle $x°$.   [3]

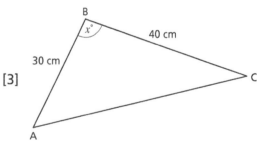

4   Find the area of the triangle.   [4]

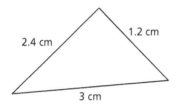

Total Marks _____ / 14

Trigonometric Identities

1. Show that $\cos^2 x - \sin^2 x = 2\cos^2 x - 1$. [2]

2. Simplify $\sin^4 x + \sin^2 x\cos^2 x$. [2]

3. Show that $\frac{1}{\cos x} - \cos x = \sin x \tan x$. [3]

4. Given that $\cos x = \frac{3}{4}$ and x is reflex:

   a) Find the value of $\sin x$. [2]

   b) Find the value of $\tan x$. [2]

5. Given that $x = \sin\theta$ and $y = \cos\theta$, eliminate θ to give an equation in terms of x and y. [2]

Total Marks _____ / 13

Solving Trigonometric Equations

1 Solve the equation $(\sin x - 1)(5\cos x + 3) = 0$ in the interval $0° \leqslant x \leqslant 360°$. [4]

2 Solve the equation $\sin^2(x - 30°) = \frac{1}{2}$, $-180° \leqslant x \leqslant 180°$. [3]

3 **a)** Show that $\cos^2 3x + 4\sin 3x - 1 = \sin 3x(4 - \sin 3x)$. [3]

b) Hence, solve $\sin 3x(4 - \sin 3x) = 0$, $0° \leqslant x \leqslant 360°$. [4]

Total Marks .... / 14

Indefinite Integrals

You must be able to:

- Know and use the fundamental theorem of calculus
- Understand that integration is the reverse of the process of differentiation
- Given f'(x) and a point on the curve, you should be able to find an equation of the curve in the form $y = f(x)$.

Integrating Functions in the Form ax^n

- Calculus is the collective term for differentiation and **integration**.
- Integration is the process of finding y when you know $\frac{dy}{dx}$.
- If $\frac{dy}{dx} = ax^n$ then $y = \frac{ax^{n+1}}{n+1} + c$.

Find y when $\frac{dy}{dx} = 2x$.

$\frac{dy}{dx} = 2x^1$ ← Remember $x = x^1$

$y = \frac{2x^{1+1}}{1+1} + c = \frac{2x^2}{2} + c$

$y = x^2 + c$

- The same rules apply when dealing with all powers.

Find y if:

a) $\frac{dy}{dx} = x^5$

$y = \frac{x^6}{6} + c$ ← Raise the power by 1 and divide by the new power. Remember to add the constant of integration.

b) $\frac{dy}{dx} = x^{-3}$

$y = \frac{x^{-2}}{-2} + c$ ← Take care when adding 1 to a negative power.

c) $\frac{dy}{dx} = 4x^7$

$y = \frac{4x^8}{8} + c$

$y = \frac{1}{2}x^8 + c$ ← Remember to simplify where possible.

d) $\frac{dy}{dx} = \frac{2}{5}x^{-\frac{1}{4}}$

$y = \frac{\frac{2}{5}x^{-\frac{1}{4}+1}}{-\frac{1}{4}+1} + c$ ← Take care when adding 1 to fractional powers.

$y = \frac{\frac{2}{5}x^{\frac{3}{4}}}{\frac{3}{4}} + c$

$y = \frac{4}{3} \times \frac{2}{5}x^{\frac{3}{4}} + c$ ← Remember when dividing fractions to multiply by the reciprocal.

$y = \frac{8}{15}x^{\frac{3}{4}} + c$

> **Key Point**
>
> If $y = x^n + c$, where c is a constant, $\frac{dy}{dx} = nx^{n-1}$.
> Therefore if $\frac{dy}{dx} = x^n$, then $y = \frac{x^{n+1}}{n+1} + c$.
> When dealing with **indefinite integrals**, always remember to add the **constant of integration**.

> **Key Point**
>
> Note that $\frac{x}{2} = \frac{1}{2}x$

Using the Integral Sign $\int ax^n dx$

- When integrating, you need to be familiar with the integral notation:

This symbol means to integrate.

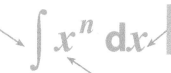

This is the letter that tells which variable you are integrating with respect to.

This is the expression to be integrated.

Key Point

$$\int \frac{dy}{dx} dx = y + c$$

$$\int x^n dx = \frac{x^{n+1}}{n+1} + c$$

1) $\int 3x \, dx$

$$\int 3x \, dx = \frac{3x^2}{2} + c$$

2) $\int (4x^5 + 6) \, dx$ ← You can integrate terms individually.

$$\int (4x^5 + 6) dx = \frac{4x^6}{6} + 6x + c$$ ← Remember $6 = 6x^0$ so $\int 6 = 6x$.

$$= \frac{2x^6}{3} + 6x + c$$ ← Always simplify when possible.

3) $\int 4t(5t + 1) \, dt$

$$\int 4t(5t + 1) \, dt = \int \left(20t^2 + 4t\right) dt$$ ← Expand the brackets first.

$$= \frac{20t^3}{3} + \frac{4t^2}{2} + c$$ ← The variable is t so integrate with respect to t.

$$= \frac{20t^3}{3} + 2t^2 + c$$

Finding the Constant of Integration

- You can find the constant of integration when you are given a point on the curve that the function passes through.

Key Point

Note that $\frac{2}{x} = 2x^{-1}$

but $\frac{1}{2x} = \frac{1}{2}x^{-1}$

The gradient of a curve is given by $f'(x) = x^2 - 3x - \frac{2}{x^2}$. The curve passes through the point (1, 1). Find the equation of the curve.

$$\frac{dy}{dx} = x^2 - 3x - \frac{2}{x^2} = x^2 - 3x - 2x^{-2}$$ ← Always rearrange before integrating.

$$\int x^2 - 3x - 2x^{-2} = \frac{x^3}{3} - \frac{3x^2}{2} - \frac{2x^{-1}}{-1} + c$$

$$y = \frac{x^3}{3} - \frac{3x^2}{2} + \frac{2}{x} + c$$

$$1 = \frac{(1)^3}{3} - \frac{3(1)^2}{2} + \frac{2}{1} + c, \quad c = \frac{1}{6}$$ ← Using the point (1, 1) when $x = 1$, $y = 1$.

$$y = \frac{x^3}{3} - \frac{3x^2}{2} + \frac{2}{x} + \frac{1}{6}$$

Quick Test

1. Find y when $\frac{dy}{dx} = 7x^2 - 5x + 2$.   2. Find $\int x^3 + \frac{1}{2x^2} \, dx$.

3. Find the equation of the curve with gradient function $f'(x) = 3x^2 + 2x$ and which passes through the point (2, 10).

Key Words

calculus
integration
indefinite integral
constant of integration

Integration to Find Areas

You must be able to:

- Evaluate definite integrals
- Use a definite integral to find the area under a curve.

Definite Integrals

- Integrating functions with defined limits is called **definite integration**.
- The **definite integral** can be defined as $\int_a^b f'(x)\,dx = \left[f(x)\right]_a^b = f(b) - f(a)$.
- For example:

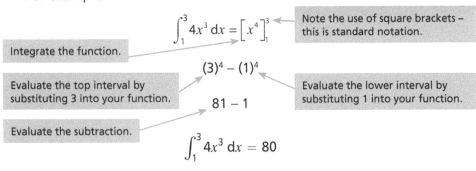

$$\int_1^3 4x^3\,dx = \left[x^4\right]_1^3$$

Integrate the function.

Note the use of square brackets – this is standard notation.

$$(3)^4 - (1)^4$$

Evaluate the top interval by substituting 3 into your function.

Evaluate the lower interval by substituting 1 into your function.

$$81 - 1$$

Evaluate the subtraction.

$$\int_1^3 4x^3\,dx = 80$$

> **Key Point**
>
> When dealing with definite integrals, you can ignore the constant of integration as this will cancel out when you complete the subtraction 'top interval'– 'bottom interval'.

Evaluate:

a) $\int_1^4 2x^3 - 4x + 5\,dx$

$$\int_1^4 2x^3 - 4x + 5\,dx = \left[\frac{2x^4}{4} - \frac{4x^2}{2} + 5x\right]_1^4$$

$$= \left[\frac{x^4}{2} - 2x^2 + 5x\right]_1^4$$

Simplify before you substitute in.

$$= \left(\frac{4^4}{2} - 2(4)^2 + 5(4)\right) - \left(\frac{1^4}{2} - 2(1)^2 + 5(1)\right)$$

$$= 116 - \frac{7}{2} = 112.5$$

> **Key Point**
>
> Check if your integration is correct by differentiating and seeing if you get back to your original function.

b) $\int_2^3 \frac{2 + \sqrt{x}}{x^2}\,dx$

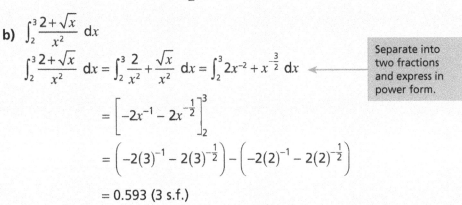

$$\int_2^3 \frac{2 + \sqrt{x}}{x^2}\,dx = \int_2^3 \frac{2}{x^2} + \frac{\sqrt{x}}{x^2}\,dx = \int_2^3 2x^{-2} + x^{-\frac{3}{2}}\,dx$$

Separate into two fractions and express in power form.

$$= \left[-2x^{-1} - 2x^{-\frac{1}{2}}\right]_2^3$$

$$= \left(-2(3)^{-1} - 2(3)^{-\frac{1}{2}}\right) - \left(-2(2)^{-1} - 2(2)^{-\frac{1}{2}}\right)$$

$$= 0.593 \text{ (3 s.f.)}$$

> **Key Point**
>
> Using function notation, the differential of f(x) is f'(x).
>
> integrate $\curvearrowleft$ f'(x) $\curvearrowright$ differentiate
>
> $\quad$ f(x)
>
> integrate $\curvearrowleft$ $\int$f(x) $\curvearrowright$ differentiate

Using Definite Integrals to Find Areas Under Curves

- The area found by evaluating $\int_a^b f(x)\,dx$ can be seen in the diagram on the right. The value of the definite integral is the area of the **region** bounded by the curve $y = f(x)$, the x-axis and the lines $x = a$ and $x = b$.

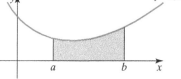

1) The diagram shows a sketch of the curve with equation

   $$y = 3x + \frac{6}{x^2}, x > 0.$$

   The region R is bounded by the curve, the x-axis and the lines $x = 3$ and $x = 1$.

   Find the area of R.

   $$\int_1^3 3x + \frac{6}{x^2}\,dx$$

   $$= \int_1^3 (3x + 6x^{-2})\,dx$$

   $$= \left[\frac{3}{2}x^2 - 6x^{-1}\right]_1^3$$

   $$= \left(\frac{3}{2}(3)^2 - 6(3)^{-1}\right) - \left(\frac{3}{2}(1)^2 - 6(1)^{-1}\right) = 16$$

> **Key Point**
>
> Evaluating the integral of a function which lies below the x-axis will give a negative answer.
>
> If the function lies above and below the x-axis, you may have to find the area by using separate intervals.

The line $x = 3$ forms the upper limit and the line $x = 1$ forms the lower limit.

2) Find the **finite** region bounded by the curve $y = x(x - 4)(x + 2)$ and the x-axis.

   When $y = 0$, $x = 0$,
   $x = 4$, $x = -2$.

    Firstly sketch the curve.

   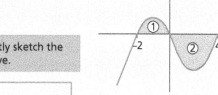

   When $x = 0$, $y = 0$.

   Region 1

   $$\int_{-2}^0 x(x - 4)(x + 2)\,dx$$

   $$\int_{-2}^0 (x^3 - 2x^2 - 8x)\,dx$$

   $$= \left[\frac{1}{4}x^4 - \frac{2}{3}x^3 - 4x^2\right]_{-2}^0$$

   $$= \left(\frac{1}{4}(0)^4 - \frac{2}{3}(0)^3 - 4(0)^2\right) - \left(\frac{1}{4}(-2)^4 - \frac{2}{3}(-2)^3 - 4(-2)^2\right) = \frac{20}{3}$$

   Region 2

   $$\left[\frac{1}{4}x^4 - \frac{2}{3}x^3 - 4x^2\right]_0^4$$

   $$= \left(\frac{1}{4}(4)^4 - \frac{2}{3}(4)^3 - 4(4)^2\right) - \left(\frac{1}{4}(0)^4 - \frac{2}{3}(0)^3 - 4(0)^2\right) = -\frac{128}{3}$$

   Total area $R = \dfrac{20}{3} + \dfrac{128}{3} = \dfrac{148}{3}$

Note the part of the area bounded by the curve, the x-axis and the lines $x = 4$ and $x = 0$ is below the x-axis, so the whole region needs to be split into two areas.

This is negative as the area is below the x-axis, so it needs to be changed to positive before being added to region 1.

Quick Test

1. a) Evaluate $\int_{-2}^5 2x^2 + 5\,dx$.   b) Evaluate $\int_1^3 \frac{x^3 + 3}{x^2}\,dx$.

2. Sketch the curve with equation $y = (x + 1)(x + 2)$. Hence find the area of the finite region bounded by the curve and the x-axis.

> **Key Words**
>
> definite integral
> region
> finite

Introduction to Vectors

You must be able to:

- Use vectors in two dimensions
- Calculate the magnitude and direction of vectors
- Add vectors and multiply vectors by scalars
- Understand and use the triangle and parallelogram laws of addition.

Introduction to Vectors

- **Vectors** are mathematical quantities having both **magnitude** (size) and **direction**. They are used to represent certain quantities in mechanics. For example, 'force' and 'acceleration' are vector quantities.
- A vector in mathematics may be represented by a directed line:

  A ——————————————→ B

  This vector may be written as $\overrightarrow{AB}$, or denoted by a **bold** letter, say **a**.
- The vector −**a** is a vector of the same magnitude as **a**, but pointing in the opposite direction. So $-\mathbf{a} = -\overrightarrow{AB} = \overrightarrow{BA}$

  A ←—————————————— B
- |**a**| indicates the magnitude of the vector **a**.

> **Key Point**
>
> When you write down vectors, you should underline the letter representing the vector, as you cannot write in 'bold'!

Vector Addition

Triangle Law of Addition

- Vectors may be added using the triangle law of addition. For example, in the diagram (right), the vector **c** = **a** + **b**. **c** is called the **resultant** of **a** and **b**.

- If you are given the magnitude and directions of both **a** and **b**, then you can find the magnitude and direction of **c** as follows:
  - Suppose |**a**| = 3 and |**b**| = 5, with the direction of **b** lying at an angle of 40° from the horizontal as in the diagram (right).
  - Then |**c**| may be found using the cosine rule:

    $$|\mathbf{c}|^2 = 3^2 + 5^2 - 2 \times 3 \times 5 \times \cos 140° \Rightarrow |\mathbf{c}| = 7.55$$

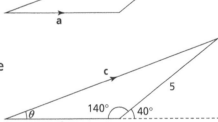

  - The direction of **c** may be found by using the sine rule:

    $$\frac{\sin \theta}{5} = \frac{\sin 140}{7.549} \Rightarrow \theta = 25.2°$$

  - So the vector **c** has magnitude 7.55, acting at 25.2° to the horizontal.

> In calculations such as this, ensure you use at least 4 significant figures here, i.e. do not use rounded values prematurely or your final answer may be incorrect.

Parallelogram Law of Addition

- The parallelogram law for the addition of vectors works in a similar way to the triangle law.
- Suppose you are given two vectors of magnitudes 6 and 14 respectively:
  - The resultant vector **R** may be found by completing a parallelogram, and applying the sine and cosine rules, similar to before.

    $$|\mathbf{R}|^2 = 6^2 + 14^2 - 2 \times 6 \times 14 \times \cos 120° \Rightarrow |\mathbf{R}| = 17.8$$

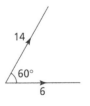

  - The direction of **R** is found by using the sine rule:

    $$\frac{\sin \theta}{14} = \frac{\sin 120}{17.8} \Rightarrow \theta = 43.0°$$

  - Therefore **R** has magnitude 17.8, acting at an angle of 43.0° to the horizontal.

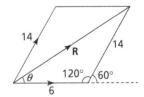

Cartesian Vector Notation

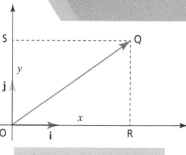

- A vector of magnitude 1 in the direction of the x-axis is denoted **i**.
- A vector of magnitude 1 in the direction of the y-axis is denoted **j**.
- Expressing $\overrightarrow{OQ}$ in Cartesian vector notation: if Q has coordinates (x, y) then the vector $\overrightarrow{OQ}$ may be written as $\overrightarrow{OQ} = x\mathbf{i} + y\mathbf{j}$.

- $\overrightarrow{OQ}$ may also be expressed by the column vector $\begin{pmatrix} x \\ y \end{pmatrix}$.

- So $\overrightarrow{OQ} = x\mathbf{i} + y\mathbf{j} = \begin{pmatrix} x \\ y \end{pmatrix}$. These are all equivalent.

Finding Magnitude and Direction

- Suppose $\overrightarrow{OQ} = x\mathbf{i} + y\mathbf{j} = \begin{pmatrix} x \\ y \end{pmatrix}$. Then $|\overrightarrow{OQ}| = \sqrt{x^2 + y^2}$. And $\theta = \tan^{-1}\left(\frac{y}{x}\right)$.

> Using Pythagoras's theorem.

> Using right-angled trigonometry.

Find the magnitude and direction of the vector $\mathbf{a} = 12\mathbf{i} + 5\mathbf{j}$.

$|\mathbf{a}| = \sqrt{12^2 + 5^2} = 13$ $\theta = \tan^{-1}\left(\frac{5}{12}\right) = 22.6°$ to the horizontal

Finding a Vector in Cartesian Notation

- You also need to be able to work in reverse, i.e. express a vector in Cartesian notation, having been given its magnitude and direction.

$$\cos\theta = \frac{x}{|\overrightarrow{OQ}|} \Rightarrow x = |\overrightarrow{OQ}|\cos\theta \qquad \sin\theta = \frac{y}{|\overrightarrow{OQ}|} \Rightarrow y = |\overrightarrow{OQ}|\sin\theta$$

Then $\overrightarrow{OQ} = (|\overrightarrow{OQ}|\cos\theta)\mathbf{i} + (|\overrightarrow{OQ}|\sin\theta)\mathbf{j}$

Suppose a vector **a** has magnitude 16 and acts in a direction of 30° to the horizontal. Express **a** in Cartesian vector form.

So $x = 16\cos 30° = 8\sqrt{3}$ $y = 16\sin 30° = 8$

Therefore $\mathbf{a} = 8\sqrt{3}\mathbf{i} + 8\mathbf{j}$.

Key Point
Cartesian notation allows you to find the magnitude and direction of a vector quickly and easily.

Application of Unit Vectors

- You need to be able to find unit vectors in the direction of a given vector. To do this, simply divide the vector by its magnitude.

Find a unit vector in the direction of $7\mathbf{i} - 24\mathbf{j}$.

Using Pythagoras, $|7\mathbf{i} - 24\mathbf{j}| = \sqrt{7^2 + 24^2} = \sqrt{625} = 25$.

> You can easily check this is a unit vector, since $\left(\frac{7}{25}\right)^2 + \left(\frac{24}{25}\right)^2 = 1$

So the required vector is $\frac{7\mathbf{i} - 24\mathbf{j}}{25}$, or $\frac{7}{25}\mathbf{i} - \frac{24}{25}\mathbf{j}$.

Key Point
You are not expected to 'learn' the general results above left, only be able to use them.

- You need to be able to find a vector of given magnitude and direction.

Find a vector of magnitude 15 in the direction $2\mathbf{i} + \mathbf{j}$.

> 15 multiplied by the unit vector.

A unit vector in this direction is $\frac{2\mathbf{i} + \mathbf{j}}{\sqrt{2^2 + 1^2}} = \frac{2\mathbf{i} + \mathbf{j}}{\sqrt{5}}$

A vector of magnitude 15 in this direction is $\frac{15(2\mathbf{i} + \mathbf{j})}{\sqrt{5}} = 6\sqrt{5}\mathbf{i} + 3\sqrt{5}\mathbf{j}$

Key Point
You should be prepared to rationalise surds where they are seen.

Quick Test

1. A vector **p** has magnitude 12 and acts in a direction of 30° from the x-axis. Express **p** in Cartesian notation.
2. Find a unit vector in the direction of $2\mathbf{i} + 4\mathbf{j}$, giving your answer in rationalised surd form.

Key Words
vector
magnitude
direction
resultant

Vector Geometry

You must be able to:

- Understand and use position vectors
- Calculate the distance between two points by using the distance formula
- Use vectors to solve problems in pure mathematics.

Position Vectors and Distance Formula

- Given a point in the plane written as a coordinate, you can also express this as a **position vector**.
- For example, consider the points A(5, 7) and B(3, −2):

  - The position vector of A is given by $\overrightarrow{OA} = \mathbf{a} = 5\mathbf{i} + 7\mathbf{j} = \begin{pmatrix} 5 \\ 7 \end{pmatrix}$.

  - Similarly, the position vector of B is given by $\overrightarrow{OB} = \mathbf{b} = 3\mathbf{i} - 2\mathbf{j} = \begin{pmatrix} 3 \\ -2 \end{pmatrix}$.

  - The position vector of B relative to A is given by $\overrightarrow{AB}$. This can also be described as the **relative position vector** of B from A.
  - From the diagram, you can see that $\overrightarrow{AB} = -\mathbf{a} + \mathbf{b}$, or written more simply as $\overrightarrow{AB} = \mathbf{b} - \mathbf{a}$.
  - In this case, $\overrightarrow{AB} = \mathbf{b} - \mathbf{a} = (3\mathbf{i} - 2\mathbf{j}) - (5\mathbf{i} + 7\mathbf{j}) = -2\mathbf{i} - 9\mathbf{j}$.

- The distance between two points (x_1, y_1) and (x_2, y_2) is given by the distance formula: $d = \sqrt{(x_2 - x_1)^2 + (y_2 - y_1)^2}$
- So referring to the points A(5, 7) and B(3, −2) once more, the distance

$$|\overrightarrow{AB}| = d = \sqrt{(3 - 5)^2 + (-2 - 7)^2} = \sqrt{4 + 81} = \sqrt{85}$$

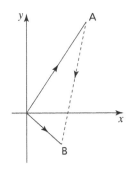

> **Key Point**
>
> You need two sets of coordinates to use the distance formula, but it does not matter which you choose to be (x_1, y_1) and which you choose to be (x_2, y_2).

Solving Problems Using Vectors

Splitting a Line in a Given Ratio

- A common type of question is having to find the position vector of a point C on a line joining A to B, where C must split AB in a given ratio.

In the diagram, A is the point (1, 8) and B is the point (2, −3).
Find the position vector of the point C that splits AB in the ratio 2 : 5.

$$\overrightarrow{OC} = \overrightarrow{OA} + \overrightarrow{AC} \qquad \overrightarrow{OA} = \begin{pmatrix} 1 \\ 8 \end{pmatrix}$$

$$\overrightarrow{AB} = -\overrightarrow{OA} + \overrightarrow{OB} = -\begin{pmatrix} 1 \\ 8 \end{pmatrix} + \begin{pmatrix} 2 \\ -3 \end{pmatrix} = \begin{pmatrix} 1 \\ -11 \end{pmatrix}$$

It is crucial that you realise the required fraction is $\frac{2}{7}$, and not $\frac{2}{5}$, which is a common mistake.

$$\overrightarrow{AC} = \frac{2}{7}\overrightarrow{AB}$$

$$= \frac{2}{7}\begin{pmatrix} 1 \\ -11 \end{pmatrix} = \begin{pmatrix} \frac{2}{7} \\ -\frac{22}{7} \end{pmatrix}$$

So $\overrightarrow{OC} = \overrightarrow{OA} + \overrightarrow{AC} = \begin{pmatrix} 1 \\ 8 \end{pmatrix} + \begin{pmatrix} \frac{2}{7} \\ -\frac{22}{7} \end{pmatrix} = \begin{pmatrix} \frac{9}{7} \\ \frac{34}{7} \end{pmatrix}$

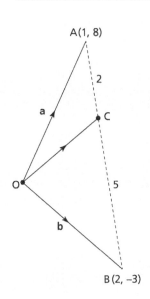

Finding a Fourth Point on a Given Shape

- You can use the **geometrical properties** of a shape if you are asked to find a missing point.

A parallelogram ABCD has coordinates A(2, 3), B(5, 7) and C(10, 5). Find the coordinates of D.

Using position vectors:

$\overrightarrow{OD} = \overrightarrow{OC} + \overrightarrow{CD}$

$\therefore \overrightarrow{OD} = \overrightarrow{OC} + \overrightarrow{BA}$

$\therefore \overrightarrow{OD} = \overrightarrow{OC} + (\overrightarrow{OA} - \overrightarrow{OB})$

$= \begin{pmatrix} 10 \\ 5 \end{pmatrix} + \begin{pmatrix} 2 \\ 3 \end{pmatrix} - \begin{pmatrix} 5 \\ 7 \end{pmatrix} = \begin{pmatrix} 7 \\ 1 \end{pmatrix}$

> Since CD and BA are parallel and have the same length, their vectors are equal.

> Using the formula for the position vector of A relative to B.

So D has coordinates (7, 1).

> **Key Point**
>
> Recall that a parallelogram is a quadrilateral where each pair of opposite sides are parallel and have the same length.

Problems Involving Simultaneous Equations

- These types of question use similar techniques to those seen at GCSE.

In the diagram, $\overrightarrow{OA} = \mathbf{a}$ and $\overrightarrow{OB} = \mathbf{b}$. C splits the line OA in the ratio 4 : 3 and E is the midpoint of AB. Find $\overrightarrow{OD}$ in terms of $\mathbf{a}$ and $\mathbf{b}$.

C, D and B are collinear, as are O, D and E. You can therefore use the facts that $\overrightarrow{CD} = \lambda\overrightarrow{CB}$ and $\overrightarrow{OD} = \mu\overrightarrow{OE}$.

$\overrightarrow{OD} = \overrightarrow{OC} + \overrightarrow{CD} = \dfrac{4}{7}\mathbf{a} + \lambda\left(-\dfrac{4}{7}\mathbf{a} + \mathbf{b}\right) = \left(\dfrac{4}{7} - \dfrac{4\lambda}{7}\right)\mathbf{a} + \lambda\mathbf{b}$

Also, $\overrightarrow{OD} = \mu\overrightarrow{OE} = \mu(\overrightarrow{OA} + \overrightarrow{AE}) = \mu\left(\mathbf{a} + \dfrac{1}{2}(-\mathbf{a} + \mathbf{b})\right) = \dfrac{\mu}{2}\mathbf{a} + \dfrac{\mu}{2}\mathbf{b}$

$\left(\dfrac{4}{7} - \dfrac{4\lambda}{7}\right) = \dfrac{\mu}{2}$

> Comparing the coefficients of **a**.

$\lambda = \dfrac{\mu}{2}$

> And comparing the coefficients of **b**.

Therefore $\dfrac{4}{7} - \dfrac{4\lambda}{7} = \lambda$.

So $\dfrac{11\lambda}{7} = \dfrac{4}{7}$ and therefore $\lambda = \dfrac{4}{11}$

So $\overrightarrow{OD} = \dfrac{4}{11}\mathbf{a} + \dfrac{4}{11}\mathbf{b}$

> **Key Point**
>
> Three points are **collinear** if they lie on the same line.

Quick Test

1. Points A, B and C have position vectors $\overrightarrow{OA} = 4\mathbf{i} - 10\mathbf{j}$, $\overrightarrow{OB} = 2\mathbf{j}$ and $\overrightarrow{OC} = 7\mathbf{i} + 2\mathbf{j}$.
   a) Find the vectors $\overrightarrow{AB}$ and $\overrightarrow{BC}$.
   b) Find the value of the length ratio $\dfrac{|\overrightarrow{AB}|}{|\overrightarrow{BC}|}$

2. A is the point (2, 11) and B is the point (16, −3). Find the position vector of the point C, which splits AB in the ratio 3 : 5.

> **Key Words**
>
> position vector
> relative position vector
> distance formula
> geometrical properties
> collinear

Proof

You must be able to:

- Construct mathematical proofs using algebra
- Use proof by exhaustion and disproof by counter-example.

Mathematical Proof

- **Proof** is a logical and structured argument to show that a mathematical statement is always true.

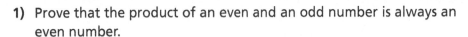

```
A conjecture or statement
        ↓
Use of known theorems/facts
        ↓
    Logical steps
        ↓
 Mathematical conclusion
```

Key Point

Show every step of your proof clearly.

Key Point

Make sure you have covered all cases.

Key Point

Ensure every step follows on from the previous one.

1) Prove that the product of an even and an odd number is always an even number.

$2n$ can be used to represent any even number, where n is an integer.

$2n + 1$ can be used to represent any odd number, where n is an integer.

$2n(2n + 1) = 4n^2 + 2n$

$\quad\quad\quad\quad\quad = 2(2n^2 + n)$, which is an even number.

Therefore the product of an odd and an even number is always an even number.

2) Prove that $3n^2 - 4n + 15$ is positive for all values of n.

$3n^2 - 4n + 15$

$= 3\left(n^2 - \dfrac{4}{3}n + 5\right)$

$= 3\left[\left(n - \dfrac{4}{6}\right)^2 - \dfrac{16}{36} + 5\right] = 3\left(n - \dfrac{2}{3}\right)^2 + \dfrac{41}{3}$ ← Complete the square.

As $\left(n - \dfrac{2}{3}\right)^2$ is a square number, it is always greater than zero.

$\therefore 3\left(n - \dfrac{2}{3}\right)^2 + \dfrac{41}{3} > 0$ and so positive for all values of n.

Key Point

State any assumptions you have used.

3) Show that the equation of the normal to the line with equation $y = x^3 - 3x + 2$ at the point (2, 4) is $9y + x = 38$.

$y = x^3 - 3x + 2$

$\dfrac{dy}{dx} = 3x^2 - 3$

At $x = 2$, $\dfrac{dy}{dx} = 3(2)^2 - 3 = 9$

Gradient of normal $= -\dfrac{1}{9}$ ← Remember that the normal is perpendicular to the tangent.

$\therefore y - 4 = -\dfrac{1}{9}(x - 2)$

$\therefore 9y + x = 38$

Therefore shown.

> **Key Point**
>
> Use any rules you know to help you with your proof.

Proof by Exhaustion

- A mathematical statement can be proved by exhaustion.
- Proof by exhaustion involves proving the statement is true for all cases.

Prove that none of the squares of the integer numbers 1 to 6 end in a 7.

$1^2 = 1,\qquad 2^2 = 4,\qquad 3^2 = 9,\qquad 4^2 = 16,\qquad 5^2 = 25,\qquad 6^2 = 36$

The list covers all possible cases and none of the cases end in a 7.

Therefore the statement has been proved by exhaustion.

Disproof by Counter-example

- Proof (or disproof) by counter-example is to find an example that shows that the statement is false.

1) It is suggested that for every prime number p, $2p + 1$ is also prime.

Give a counter-example to disprove this statement.

If $p = 7$, $2p + 1 = 15$.

15 is not a prime number.

This counter-example disproves the statement.

2) Give a counter-example for the following statement:

If x and y are irrational real numbers, then xy is also irrational.

If $x = \sqrt{8}$ (irrational) and $y = \sqrt{2}$ (irrational)

$xy = \sqrt{8 \times 2} = \sqrt{16} = 4$

4 is rational and so this counter-example disproves the statement.

> **Quick Test**
>
> 1. Prove that $(a + b)^2 = a^2 + 2ab + b^2$.
> 2. Prove that the product of two odd numbers is always an odd number.
> 3. Prove that $y = x^2 - 3x + 15$ has no real roots.

> **Key Words**
>
> proof

Indefinite Integrals

1 Evaluate:

    a) $\int (x^4 + 3x)\,dx$ [1]

    b) $\int (2t^{\frac{1}{2}} + t)\,dt$ [1]

    c) $\int (x+1)(x+2)\,dx$ [1]

2 Evaluate $\int \dfrac{6x-2}{3x^3}\,dx$.   [4]

3 **a)** Show that $\left(2 - 3\sqrt{x}\right)^2$ can be written in the form $4 - a\sqrt{x} + bx$, where a and b are constants to be found.   [2]

    b) Hence find $\int \left(2 - 3\sqrt{x}\right)^2\,dx$.   [3]

4 A curve passes through the point (3, −1) and the gradient function is $f'(x) = 2x + 5$. Find the equation of the curve.   [3]

5 Find s in terms of t if $\dfrac{ds}{dt} = 3t - \dfrac{8}{t^2}$, given that when $s = \dfrac{3}{2}$, $t = 1$.   [2]

> **Total Marks** .... / 17

Integration to Find Areas

1 Calculate the following integrals:

    a) $\displaystyle\int_{1}^{6} (5x+1)\,dx$ [2]

    b) $\displaystyle\int_{20}^{50} (-q^3 + 100q)\,dq$ [2]

    c) $\displaystyle\int_{2}^{3} \left(\dfrac{x-1}{x^3}\right)\,dx$ [2]

2   Find the area of the finite region bounded by the curve with equation $y = 3x^2 - 2x + 2$, the x-axis and the lines $x = 2$ and $x = 0$.   [4]

3   Given that $\int_{-5}^{9} t(x)dx = -4$, $\int_{6}^{9} t(x)dx = 4$ and $\int_{-3}^{6} t(x)dx = -9$,

find $\int_{-5}^{-3} t(x)dx$.   [3]

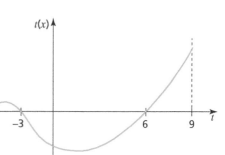

4   A curve C has equation $y = 15x^{\frac{1}{2}} + x^{\frac{3}{2}}$

Use calculus to find the region bounded by the curve, the x-axis and the lines $x = 1$ and $x = 2$.

Give your answer to 3 significant figures.   [4]

Total Marks .... / 17

Introduction to Vectors

1   Find a vector of magnitude 5 units in the direction of $7\mathbf{i} + 24\mathbf{j}$.   [4]

2   Express vector **a** in Cartesian form if:

a)   it has magnitude 4 and acts in a direction of 60° to the horizontal   [2]

b)   it has magnitude 10 and acts in a direction of 150° to the horizontal.   [2]

3 Find the magnitude and direction of the resultant of vectors **a** and **b**: [6]

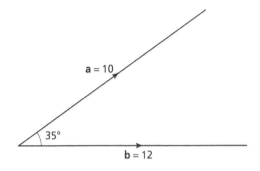

Total Marks _____ / 14

Vector Geometry

1 The distance between the points (2, 3) and (7, k) is $\sqrt{30}$.

Find the possible values for k. [4]

2 A is the point (−30, 10) and B is the point (45, 80).

Find the position vector of the point C which splits AB in the ratio 3 : 7. [6]

3 A parallelogram OABC has vectors $\overrightarrow{OA}$ = **a** and $\overrightarrow{OC}$ = **c**.

Prove that the intersection of the diagonals of the parallelogram bisects each diagonal. [10]

Total Marks _____ / 20

Proof

1 Prove that the square of any odd number is always one more than a multiple of 4. [3]

2 Write down the nth term of the sequence 4, 7, 10, 13, 16, …

Prove that the product of any two terms of this sequence is also a term of the sequence. [3]

3 ABC is a triangle. P is a point on AB such that AP = PC = BC. Angle BAC = x.

Prove that angle ABC = $2x$. [3]

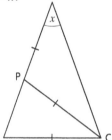

4 Show that $(2a - 1)^2 - (2b - 1)^2 = 4(a - b)(a + b - 1)$. [3]

5 Prove that $(3n + 1)^2 - (3n - 1)^2$ is a multiple of 4, for all positive integer values of n. [3]

6 Prove that the quadrilateral A(1, 1), B(2, 4), C(6, 5) and D(5, 2) is a parallelogram. [3]

Total Marks _____ / 18

Review Questions

Exponentials and Logarithms

1 Sketch the graph of $y = 2^{x-2}$, showing clearly any asymptotes and intersections with any axes. [3]

2 Solve the equation $3\ln(2x - 3) = 1$. [3]

3 Find $\dfrac{dy}{dx}$, given $y = (e^{3x} - 2)^2$. [4]

Total Marks _____ / 10

Laws of Logarithms

1 Solve the equation $2^{x-1} = \dfrac{1}{5}$, giving your answer to 3 significant figures. [3]

2 Solve the equation $\log_{10}5 + 2\log_{10}x = 1 + \log_{10}8$. [6]

3. Write $\log_{10} \sqrt{p^4 q^7}$ in the form $a\log_{10} p + b\log_{10} q$. [4]

Total Marks _____ / 13

Modelling Using Exponentials and Logarithms

1. The amount of radioactive substance, N, is related to time, t, by the equation $N = N_0 e^{-\lambda t}$.
The half-life of a radioactive substance is the time taken for the amount of radioactive substance to halve.

[4]

Show that the half-life $t_{\frac{1}{2}} = \dfrac{\ln 2}{\lambda}$

2. It is suggested that two variables, x and y, are related by the equation $y = kb^x$. A student plots a graph of $\log_{10} y$ against x. By drawing a line of best fit, he estimates the gradient to be 0.40 and the y-intercept of the line to be at 1.18.

From these results, determine the likely values of k and b. [6]

3. It is suggested that two variables, x and y, are related by the equation $y = Axe^{kx}$. Given that a student has a set of paired values for x and y, what must she plot in order to determine the value of the constant A and the value of the constant k? [5]

Total Marks _____ / 15

Differential of a Polynomial

1 The function $f(x) = x^3 + 2x^2 + 2x$.

    a) Find $f'(x)$. [2]

    b) Solve $f'(x) = 1$. [2]

2   **a)** Differentiate $y = 5x^{-1} + \frac{3}{2}x^{\frac{1}{3}}$ with respect to x. [2]

    b) Find the value of $\frac{dy}{dx}$ when $x = 2$. [2]

3 Find $\frac{dy}{dx}$ and $\frac{d^2y}{dx^2}$ when $y = (2x - 4)(x + 2)$. [5]

Total Marks _____ / 13

Stationary Points, Tangents and Normals

1 A curve has equation $y = 2x^3 - 8x^2 + 8x$.

   a) Find the coordinates of the turning points. [3]

   b) Determine the nature of the turning points. [2]

2 A curve is given by the equation $y = 3x^2 + 3x + \dfrac{1}{x^2}$, where $x > 0$. At the points A, B and C on the curve, $x = 1$, 2 and 3 respectively.

   Find the value of the gradient at A, B and C. [4]

3 A curve C has equation $x^3 - 5x^2 + 5x + 2$.

   a) Find $\dfrac{\mathrm{d}y}{\mathrm{d}x}$ [1]

   b) The points P and Q lie on C. The gradient at both P and Q is 2. The x-coordinate of P is 3.

   i) Find the x-coordinate of Q. [1]

   ii) Find the equation of the tangent to C at P, giving your answer in the form $ax + by + c = 0$. [2]

   iii) Find the equation of the normal at the point Q. [2]

   iv) If this tangent intersects the coordinate axes at the points R and S, find the length of RS, giving your answer as a surd. [2]

Total Marks _____ / 17

Populations and Samples

You must be able to:

- Understand and use the terms 'population' and 'sample'
- Understand and critique simple random sampling, stratified sampling, systematic sampling, quota sampling and opportunity (or convenience) sampling
- Comment on the advantages and disadvantages associated with a census and a sample.

Populations, Censuses and Samples

- A **population** is a collection of individuals or items.
- A population could be of finite size, i.e. the number of pupils in a school.
- A population could be considered as infinite in size, i.e. it is impossible to know exactly how many people or items there are in the population.
- Where information is obtained from all members of the population, this is called a **census**.
- Where information is obtained from a sub-set of the population, this is called a **sample**.

	Advantages	Disadvantages
Census	• Every member of the population is used • It is unbiased • It gives a complete picture	• Time consuming • Expensive • Often impractical to ensure all the population is used
Sample	• Often cheaper • Sensible when testing items to destruction • Data is more easily available	• Possible introduction of **bias** • Natural variations between samples can make inference about the population uncertain

A shop manager wants to find out whether his customers are satisfied with the range of products available in the shop.

a) Give reasons why the manager should conduct a sample rather than a census.

The manager should conduct a sample because it will be difficult to ask all his customers within a reasonable time frame, so a sample will be quicker and easier.

b) Describe the sample units for the sample survey.

The sample units are the customers.

c) Discuss the disadvantages of the manager using a sample survey.

The results of the survey could be biased as he has not asked all his customers.

> **Key Point**
>
> Keep your answers in context to the question.

Sampling Techniques

- To select a sample, you may need to identify the **sampling frame**. The sampling frame is a list identifying every single sampling unit which could be included in the sample.

Simple Random Sample

- A sample of size n is called a simple random sample if every item in the population has an equal chance of being selected.
- A common technique for selecting a simple random sample is random number sampling:
  - Number each item in the sampling frame.
  - Select n random numbers, where n is sample size, using a random number generator or a random number table.
  - Select the items from the sampling frame which correspond to those random numbers and survey those items.

> **Key Point**
>
> If the same random number is generated more than once, we usually choose to ignore it. This is called sampling without replacement.

Systematic Sampling

- A sample is chosen by selecting items at regular intervals from an ordered list.
  - Number every item in the sampling frame.
  - Sample every kth element where $k = \dfrac{\text{population size}}{\text{sample size}}$
  - To find the starting point, select your first item by choosing a random number between 1 and k.

1 Quota sampling

- The population is divided into strata. The number of items in each stratum is set to try to reflect its proportion within the population. An interviewer selects the items (the process is not random).

2 Opportunity (or convenience sampling)

- Involves sampling items of the population which are conveniently placed to take part in the survey and fit the required criteria.

3 Stratified sampling

- The population is divided into strata. The number of items in each stratum is set to reflect its proportion within the population. The items within each stratum are selected at random.

	Advantages	Disadvantages
Simple random	• Simple and cheap • Minimises sample bias	• A sampling frame required • Not suitable for large populations
Systematic	• Suitable for large samples • Simple	• Can introduce bias • Only random if sample frame list is randomised
Quota	• Enables fieldwork to be done quickly • Easy administration	• Not random so introduction of sample bias • Interviewer chooses responders so interviewer bias
Opportunity	• Convenient to administer	• May not represent the population
Stratified	• Reflects the structure of the population • Ensures all strata of the population are represented	• Within the strata the problems are the same as simple random sampling • A sampling frame is required • Difficult to place items into a single stratum

A factory manager wants to survey workers about the staff facilities. He will do a stratified sample of 60 workers of different ages.

There are 80 workers aged between 18 and 30; 62 workers aged between 31 and 45; and 58 workers aged between 46 and 65.

Work out how many members of each stratum he needs to sample.

Age	Population size	Sample size
18–30	80	$\dfrac{80}{200} \times 60 = 24$
31–45	62	$\dfrac{62}{200} \times 60 = 18.6 \therefore 19$
46–65	58	$\dfrac{58}{200} \times 60 = 17.4 \therefore 17$
Total	200	$24 + 19 + 17 = 60$

Key Point

Systematic sampling is a good choice of sampling method for processes such as quality control on a production line.

Key Point

Strata are groups in which the population can be divided that are mutually exclusive (i.e. no item can be placed in two or more groups). For example, gender, age, religion, etc.

Key Point

To find the sample size for each strata:

$\dfrac{\text{strata size}}{\text{population}} \times \text{sample size}$

Key Point

Sampling bias occurs when a sample is collected in such a way that some members of the population are less likely to be included than others.

Key Words

population
census
sample
bias
sampling frame
strata

Quick Test

1. Give one advantage and one disadvantage of using a sample rather than a census.
2. A gym has 400 members. Explain how you could select a simple random sample of 50 clients.

Measures of Central Tendency and Spread

You must be able to:

- Calculate the mean, median and mode
- Calculate the range and interpercentile ranges, variance and standard deviation.

Measures of Central Tendency

- There are three main measures of central tendency (also known as the average): the **mean**, the **median** and the **mode**.

Discrete data	Continuous data
Can only take specific values in a given range	Can take any value in a given range

1) Ten randomly chosen children were asked how many pets they have. Their responses were: 1 3 4 5 2 3 2 2 4 2

Find the value of the mean, median and mode.

> This data is listed so the mean = $\frac{\sum x}{n}$ formula is used.

$$\text{Mean} = \frac{\sum x}{n} = \frac{1+3+4+5+2+3+2+2+4+2}{10} = \frac{28}{10} = 2.8$$

Ordering the data gives 1 2 2 2 2 3 3 4 4 5

> To find the median, the data must be in order.

The median is $\frac{1}{2}(2+3) = 2.5$

> 10 observations, so the median is halfway between the fifth and sixth observations.

The mode is 2 as this is the value which occurs most often.

2) Esmai recorded the shoe size, x, of the pupils in her year. The results are shown in the table (see the first two columns below). Find the value of the mean, median and mode.

Shoe size (x)	Frequency (f)	fx	Cumulative frequency
3	3	$3 \times 3 = 9$	3
4	17	$4 \times 17 = 68$	20
5	29	$5 \times 29 = 145$	49
6	20	$6 \times 20 = 120$	69
7	5	$7 \times 5 = 35$	74
Total	$\sum f = 74$	$\sum fx = 377$	

> To find the mean, add an fx column to the table given. To find the median, add a **cumulative frequency** column to your table.

The mean = $\frac{\sum fx}{\sum f} = \frac{377}{74} = 5.09$ (3 s.f.)

The mode is 5.

> This data is a frequency table so the mean = $\frac{\sum fx}{\sum f}$ formula is used.

Median: $\frac{74}{2} = 37 \therefore$ 37.5th value

> Note that this data is discrete.

Using the cumulative frequency column, the 37.5th value is 5.

> The 37th and 38th values are both 5, therefore the median is 5.

Key Point

When dealing with discrete data, calculate $\frac{n}{2}$ for the median. If $\frac{n}{2}$ is a whole number, add $\frac{1}{2}$ and pick the corresponding term. If $\frac{n}{2}$ is not a whole number, round up and pick the corresponding term.

3) A company recorded the time of a train from Bookham to London over a certain period. Times were recorded to the nearest minute and are shown in the table (see the first two columns below).

a) Estimate the value of the mean.

Time in minutes	Frequency (f)	Midpoint (x)	fx	Cumulative frequency
32–35	5	33.5	$33.5 \times 5 = 167.5$	5
36–40	10	38	$38 \times 10 = 380$	15
41–45	35	43	$43 \times 35 = 1505$	50
46–50	10	48	$48 \times 10 = 480$	60
Total	$\sum f = 60$		$\sum fx = 2532.5$	

$$\text{Mean} \frac{\sum fx}{\sum f} = \frac{2532.5}{60} = 42.2 \text{ (3 s.f.)}$$

> As the data is grouped, you can only estimate the mean by using the midpoint of the group as the value of x.

b) Find the modal class.

The modal class is 41–45. ←

As the answer is a group, it is called the modal class.

c) Estimate the value of the median.

The median is the $\frac{60}{2}$ = 30th value. ←

The 30th value is in the group 41–45.

As the data is continuous, there is **no** need to add $\frac{1}{2}$.

Using interpolation to estimate the median:

$\frac{m - 40.5}{45.5 - 40.5} = \frac{30 - 15}{50 - 15}$

$\therefore m = 42.6$ (3 s.f.)

Cumulative frequency up to 40

Cumulative frequency up to 46

As this is continuous data, the class boundaries are 40.5–45.5

Measures of Spread

- The **range** is the difference between the largest and the smallest value in the data set.
- The **interquartile range** (IQR) is the difference between the upper quartile (Q_3) and the lower quartile (Q_1).
- The **variance** is the average of the squared differences from the mean.
- The **standard deviation** is the square root of the variance.

Using the data about train journeys on page 88, calculate the standard deviation and the interquartile range.

Time in minutes	Frequency (f)	Midpoint (x)	fx	fx^2	Cumulative frequency
32–35	5	33.5	$33.5 \times 5 = 167.5$	$33.5^2 \times 5 = 5611.25$	5
36–40	10	38	$38 \times 10 = 380$	$38^2 \times 10 = 14440$	15
41–45	35	43	$43 \times 35 = 1505$	$43^2 \times 35 = 64715$	50
46–50	10	48	$48 \times 10 = 480$	$48^2 \times 10 = 23040$	60
Total	$\Sigma f = 60$		$\Sigma fx = 2532.5$	$\Sigma fx^2 = 107806.25$	

The standard deviation is $\sqrt{\frac{\Sigma fx^2}{\Sigma f} - \left(\frac{\Sigma fx}{\Sigma f}\right)^2} = \sqrt{\frac{107806.25}{60} - \left(\frac{2532.5}{60}\right)^2} = 3.90$ (3 s.f.)

$Q_1 = \frac{60}{4}$th value. The 15th value is top of the group 36–40, so 40.5.

$Q_3 = \frac{3 \times 60}{4}$th value. 45th value. In the group 41–45, so 40.5–45.5.

$\frac{Q_3 - 40.5}{45.5 - 40.5} = \frac{45 - 15}{50 - 15}$ $\therefore Q_3 = 44.8$ (3 s.f.) $\therefore$ IQR = 44.8 – 40.5 = 4.3

Key Point

Interpolation is based on proportion.

Key Point

Q_1 is the 25th percentile.

For the 80th percentile, find the $\frac{80 \times n}{100}$ th value.

The interpercentile range is the difference between the values for two percentiles.

Key Point

Variance

$\frac{\Sigma f(x - \bar{x})^2}{\Sigma f} = \frac{\Sigma fx^2}{\Sigma f} - \left(\frac{\Sigma fx}{\Sigma f}\right)^2$

'the mean of the squares' – 'the square of the mean'

$s_{xx} = \Sigma(x - \bar{x})^2$ is a summary statistic.

Quick Test

1. Find the mean, median and mode in the data: 3 4 6 2 8 8 5
2. Find the median and interquartile range for the data in this table.

	6	7	8	9
Frequency	10	20	36	15

3. Estimate the mean and standard deviation for the data in this table.

	15–20	21–25	26–30	31–39	40–50
Frequency	12	15	25	17	2

Key Words

mean
median
mode
discrete data
continuous data
cumulative frequency
interpolation
range
interquartile range
variance
standard deviation

Displaying and Interpreting Data 1

You must be able to:

- Construct and interpret a variety of diagrams including frequency polygon, histogram and scatter diagram
- Interpret regression lines and comment on correlation.

Cumulative Frequency Curves

- When data is in grouped form, you may have to estimate the median and the quartiles. You can do this with a **cumulative frequency curve**.

Key Point

Q_2 is the median.

The table shows the lengths of 60 hedgehogs.

Length of hedgehog (cm)	27–28	29–30	31–32	33–34
Frequency	12	20	25	3

Draw a cumulative frequency curve and use the curve to find estimates for the median and the interquartile range.

Length (cm)	Frequency	Cumulative frequency
27–28	12	12
29–30	20	32
31–32	25	57
33–34	3	60

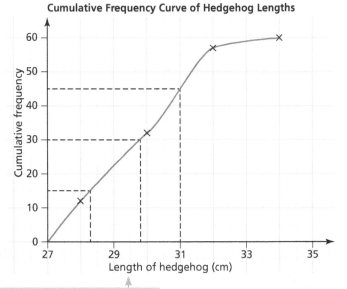

Cumulative Frequency Curve of Hedgehog Lengths

When plotting a cumulative frequency curve, the x-value is always the end of the group.

Draw lines across from 15, 30 and 45 on the cumulative frequency axis to the curve and then down to the length axis.

$$Q_1 = \frac{60}{4} = 15\text{th value} \qquad Q_2 = \frac{60}{2} = 30\text{th value} \qquad Q_3 = \frac{3 \times 60}{4} = 45\text{th value}$$

So the median is approximately 29.8 cm and the IQR is 31 − 28.3 = 2.7 cm

Histograms and Frequency Polygons

- Grouped continuous data can be displayed in a histogram.
- The principle of a histogram is the area of the bar is proportional to the frequency. The area of the bar is $k \times$ frequency.
- The height of the bar is called the frequency density.

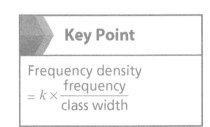

Key Point

Frequency density
$$= k \times \frac{\text{frequency}}{\text{class width}}$$

This table summarises the birth weight of babies born in a local hospital.

Weight (lbs)	5–	5.5–	6–	6.5–	7–	7.5–
Frequency	12	17	23	14	10	9

Represent this data by a histogram and estimate the number of babies who weighed between 6.2 and 7.4 lbs.

Weight	Frequency	Class width	Frequency density
5–	12	0.5	$\frac{12}{0.5} = 24$
5.5–	17	0.5	$\frac{17}{0.5} = 34$
6–	23	0.5	$\frac{23}{0.5} = 46$
6.5–	14	0.5	$\frac{14}{0.5} = 28$
7–	10	0.5	$\frac{10}{0.5} = 20$
7.5–	9	0.5	$\frac{9}{0.5} = 18$

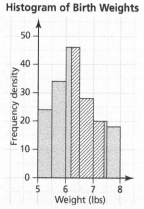

Histogram of Birth Weights

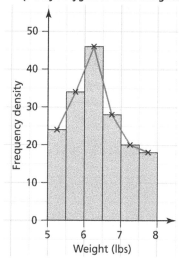

Frequency Polygon of Birth Weights

$0.3 \times 46 + 14 + 0.4 \times 20 = 36$ ← To estimate the number of babies between 6.2 and 7.4 lbs, find the area between 6.2 and 7.4

So approximately 36 babies weighed between 6.2 and 7.4 lbs.

- Joining the middle of the top of each bar forms a **frequency polygon**.

Scatter Diagrams

- Bivariate data can be represented on a scatter diagram. Correlation describes the nature of the linear relationship between two variables.
- The regression line is the line of best fit and the regression line of y on x is written in the form $y = a + bx$.

A spring has weight (g) added to it and the length of the spring is measured. The results are shown on the scatter diagram (right). The regression line of y on x has equation $y = 16 + 0.7x$.

a) Describe the correlation.

The scatter diagram shows a strong positive correlation. As the weight increases, the length of the spring increases.

b) Give a practical interpretation of the constant, a.

The spring is 16 cm long before any weight is added.

c) Give a practical interpretation of the constant, b.

The increase in the length of the spring per gram of weight added.

d) Estimate the length of the spring if 20 g is added and comment on the reliability of this estimate.

$16 + 0.7 \times 20 = 30$ cm

The estimate is not reliable as 20 g is far beyond the data range shown on the graph. ← This is extrapolation.

> ## Key Point
>
> Two variables which have correlation may not have a **causal** relationship.

Change in Length of Spring with Weight

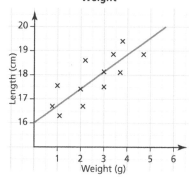

> ## Key Point
>
> For the regression line $y = a + bx$, b is the gradient and a is the y-intercept.

Quick Test

1. Plot a cumulative frequency curve and use it to estimate the IQR of the data shown right.
2. Using the table, construct a histogram and estimate the number of students who weigh between 66 kg and 78 kg.
3. The equation of regression between the length (mm) of a rod (l) and the temperature (degrees) (t) as it is heated is $l = 0.009t - 0.25$. Give a practical interpretation of the value 0.009.

Student Weights (kg)	Frequency
60–64	10
65–69	22
70–74	27
75–79	24
80–84	10
85–89	6

> ## Key Words
>
> histogram
> frequency density
> scatter diagram
> correlation
> linear
> causal

Displaying and Interpreting Data 2

You must be able to:

- Recognise and interpret possible outliers in data sets and statistical diagrams
- Construct and interpret box and whisker plots
- Select or critique data presentation techniques in the context of a statistical problem
- Understand the practicalities of dealing with large data
- Clean data, including dealing with missing data, errors and outliers
- Understand and use coding.

Outliers and Box and Whisker Diagrams

- An **outlier** is a value which lies outside the overall pattern of the rest of the data. The most common way to identify outliers is to find an upper and lower boundary using the following:

Upper boundary = $Q_3 + 1.5(IQR)$ Lower boundary = $Q_1 - 1.5(IQR)$

- Any value outside that range is considered an outlier.
- A **box and whisker diagram** displays the main features of the data, including the median and quartiles.

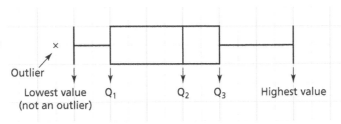

Outlier

Lowest value (not an outlier) Q_1 Q_2 Q_3 Highest value

> **Key Point**
>
> There are many ways to find outliers, so it is important to read the question to confirm which method you are required to use.

> **Key Point**
>
> When interpreting data, you will often be required to make comparisons between data sets. You should comment on a measure of spread and a measure of location.

Thirty students sat a test, which was out of 60, and the results were:

Lower quartile = 36; Upper quartile = 48; Median = 44; Lowest value = 6; Highest value = 60.

An outlier is defined as any value above $Q_3 + 1.5(IQR)$ and any value below $Q_1 - 1.5(IQR)$.

a) Given there is only one outlier, show that 6 is an outlier.

Upper outlier boundary = $48 + 1.5 \times (48 - 36) = 66$ (no outliers)

Lower outlier boundary = $36 - 1.5 \times (48 - 36) = 18$

6 < 18, so 6 is an outlier.

b) Draw a box and whisker plot to represent this data.

Box and Whisker Diagram of Test Marks

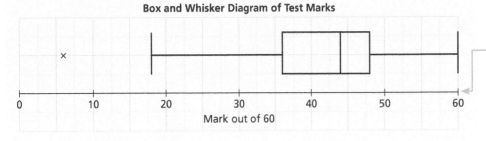

Mark out of 60

> The end of the whisker can be placed either at the lower outlier boundary or at the lowest value which is not an outlier.

Large Data Sets

- On its website, Edexcel has provided a large data set from the Met Office which you should study before the exam. You do not need to learn the data as appropriate extracts will be provided. You should, however, be familiar with the context and terminology. The data is for different variables at five UK and three overseas weather stations.

> **Key Point**
>
> You need to understand the geography associated with the data set. **n/a** means data not available. **tr** means trace amount and can be treated as zero.

- You will be required to take a sample from a large data set, identify different types of data and calculate statistics from the data.

a) Work out the mean daily mean pressure for the first 10 days in May, 1987.

$$\frac{1013 + 1008 + 1026 + 1033 + 1033 + 1033 + 1030 + 1024 + 1015 + 1014}{10} = 1022.9\,\text{hPa}$$

b) Work out the range of the mean daily wind direction.

$350 - 20 = 330°$

c) The mean daily mean pressure for Hurn is 1008.7 hPa. Meg states that the more northerly counties experience higher daily mean pressure. State with a reason whether your answer to a) supports this statement.

Leeming is north of Hurn (also in the UK) and 1022.9 > 1008.7, which might suggest Meg is correct. However, this is a very small sample size from a single location north and south, so there is not enough evidence to support Meg's statement.

LEEMING	© Crown Copyright Met Office 2015			
Date	Daily Mean Total Cloud (oktas)	Daily Mean Visibility (Dm)	Daily Mean Pressure (hPa)	Daily Mean Wind Direction (°)
01/05/1987	6	2300	1013	190
02/05/1987	5	3600	1008	270
03/05/1987	4	3100	1026	350
04/05/1987	4	4600	1033	310
05/05/1987	4	2900	1033	300
06/05/1987	3	1500	1033	20
07/05/1987	3	1700	1030	160
08/05/1987	4	1300	1024	160
09/05/1987	6	2300	1015	300
10/05/1987	6	6000	1014	290

Coding

- **Coding** is used to simplify calculations. One or more sets of data is coded to make a new set of values which are easier to work with.
- Coding will **usually** take the form $p = \frac{x - a}{b}$, where a and b are constants chosen by you or given in the question.

Coding used	Original data	Coded data
$p = \frac{x - a}{b}$	Mean $= \bar{x}$	Mean $\bar{P} = \frac{\bar{x} - a}{b}$
$p = \frac{x - a}{b}$	Standard deviation $= \sigma_x$	Standard deviation $= \sigma_p = \frac{\sigma_x}{b}$

The data here shows length (x):   110   120   130   140   150

a) Use the coding $p = \frac{x - 100}{10}$ to code the data above.

$$\frac{110 - 100}{10} \quad \frac{120 - 100}{10} \quad \frac{130 - 100}{10} \quad \frac{140 - 100}{10} \quad \frac{150 - 100}{10}$$

The coded data is: 1   2   3   4   5

b) Find the mean of the coded data.

$$\frac{1 + 2 + 3 + 4 + 5}{5} = 3$$

c) Find the mean of the original data.

$3 \times 10 + 100 = 130$

> You need to reverse the whole process when calculating the mean.

Quick Test

1. A set of data is coded using $p = \frac{v - 200}{30}$. The standard deviation of the coded data is 1.2. Find the standard deviation of the original data.
2. Using the box and whisker diagram below, write down:
   a) the median   b) the IQR   c) the value of any outliers.

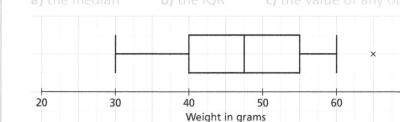

Weight in grams

Key Words
outlier
box and whisker diagram
coding

Calculating Probabilities

You must be able to:

- Calculate probabilities for single events
- Draw and interpret Venn and tree diagrams
- Understand mutually exclusive and independent events.

Calculating Probabilities for Single Events

- The probability is the chance of an event occurring.

A fair six-sided dice, numbered 1 to 6, is rolled and a fair spinner, numbered 1 to 4, is spun.

The difference in the numbers is recorded.

Find the probability that the difference is 1.

Spinner \ Dice	1	2	3	4	5	6
4	3	2	1	0	1	2
3	2	1	0	1	2	3
2	1	0	1	2	3	4
1	0	1	2	3	4	5

← Construct a sample space diagram.

← There are 24 possible outcomes.

There are seven 1s out of 24 outcomes so P(difference 1) = $\frac{7}{24}$.

> **Key Point**
>
> The **sample space** is the set of all possible outcomes.

Mutually Exclusive and Independent Events

- For **mutually exclusive** events A and B, $P(A \cap B) = 0$, i.e. the two events do not occur at the same time.
- **Independent events** have no effect on each other, e.g. rolling a dice and picking a card from a pack.
- For independent events A and B, $P(A \cap B) = P(A) \times P(B)$

> **Key Point**
>
> The rectangle in a Venn diagram represents the sample space and therefore all outcomes must be included within the rectangle.

Venn Diagrams

- **Venn diagrams** are a useful graphical representation of probabilities.

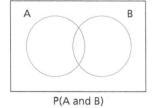
P(A and B)

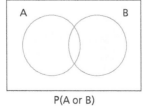
P(A or B)

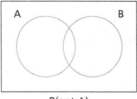
P(not A)

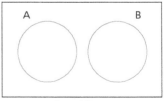
A and B are mutually exclusive

There are 40 pupils in a class: 20 play the piano, 10 play the guitar and 5 play both the piano and guitar.

a) Draw a Venn diagram to represent this information.

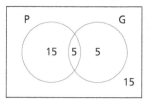

20 play the piano in total and 5 are already accounted for in the middle so 20 – 5 = 15

10 play the guitar in total and 5 are already accounted for in the middle so 10 – 5 = 5

b) A pupil is chosen at random. Find the probability that the pupil plays the piano but not the guitar.

P(play piano, not guitar) = $\frac{15}{40} = \frac{3}{8}$

15 + 5 + 5 = 25 and 40 – 25 = 15. There are 15 pupils who play neither instrument and are placed outside the circles.

This pupil should be inside the piano circle but not in the guitar circle.

Tree Diagrams

- A **tree diagram** is used to represent successive events.

A bag contains 14 red beads and 6 orange beads.

A bead is taken at random from the bag and not replaced.

A second bead is taken at random from the bag.

Find the probability that the two beads are the same colour.

To find the probability the beads are the same colour, there are two options: orange followed by orange, and red followed by red.

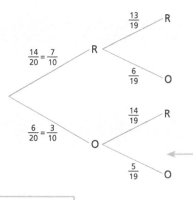

P(orange and orange) = $\frac{3}{10} \times \frac{5}{19} = \frac{3}{38}$

P(red and red) = $\frac{7}{10} \times \frac{13}{19} = \frac{91}{190}$

P(both the same colour) = $\frac{3}{38} + \frac{91}{190}$

= 0.558 (3 s.f.)

> **Key Point**
>
> When constructing a Venn diagram, always work from the inside out, i.e. fill in the centre first.

> **Key Point**
>
> Follow a pathway along the tree branches and multiply as you go.

The first bead is not replaced and so the calculations for the second branch of the tree are out of 19.

To get the total probability, add the probability of the two options together.

> **Quick Test**
>
> 1. Events A and B are mutually exclusive. P(A) = 0.2 and P(B) = 0.4.
>   a) Draw a Venn diagram to represent these two events.
>   b) Write down the probability that neither A nor B occurs.
> 2. Two fair six-sided dice are rolled and the sum of their scores recorded.
>   a) Construct a sample space diagram to represent the possible outcomes.
>   b) Find the probability of the sum being greater than or equal to 7.
> 3. A bag contains 6 red beads and 5 blue beads. A bead is drawn at random from the bag and is replaced. A second bead is then drawn from the bag.
>   Find the probability the beads are different colours.

> **Key Words**
>
> sample space
> mutually exclusive
> independent event
> Venn diagram
> tree diagram

Statistical Distributions

You must be able to:

- Understand and use discrete probability distributions
- Understand and use the binomial distribution to calculate probabilities.

Probability Distributions

- A random variable is the numerical outcome of a random event.
- A discrete variable can only take certain numerical values.
- A **probability distribution** describes all possible outcomes and their associated probabilities.

Key Point

A probability distribution must include all possible outcomes.

1) Three fair coins are tossed and the number of tails, X, is recorded.

a) Write down the probability distribution.

Outcome	Number of tails (X)
TTT	3
TTH	2
THT	2
HTT	2
HHT	1
HTH	1
THH	1
HHH	0

Construct a sample space diagram first to find all possible outcomes.

Key Point

Random variables are represented by uppercase letters, i.e. X and Y.

Lowercase letters are used when the random variable takes a particular value.

x	0	1	2	3
$P(X = x)$	$(0.5)^3 = \dfrac{1}{8}$	$(0.5)^3 \times 3$ $= \dfrac{3}{8}$	$(0.5)^3 \times 3$ $= \dfrac{3}{8}$	$(0.5)^3 = \dfrac{1}{8}$

b) Find $P(X \geqslant 1)$.

$P(X \geqslant 1) = P(X = 1) + P(X = 2) + P(X = 3) = \dfrac{7}{8}$

or $P(X \geqslant 1) = 1 - P(X = 0) = \dfrac{7}{8}$

2) A discrete random variable X has this probability distribution.
Find the value of k.

x	1	2	3	4
$P(X = x)$	k	$\dfrac{k}{2}$	$\dfrac{k}{3}$	$\dfrac{k}{4}$

$k + \dfrac{k}{2} + \dfrac{k}{3} + \dfrac{k}{4} = \dfrac{12k + 6k + 4k + 3k}{12} = \dfrac{25k}{12} = 1$ ← The sum of the probabilities is 1.

$\dfrac{25k}{12} = 1$, so $25k = 12$ ∴ $k = \dfrac{12}{25}$

Binomial Distribution

- A **binomial distribution** is a particular probability model which can be used in the circumstances shown in the table (right).

- If a random variable X can be modelled by a binomial distribution B(n, p), then P$(X = r) = \binom{n}{r} p^r(1 - p)^{n-r}$

- The binomial distribution is used for discrete random variables.

Conditions for Binomial Distribution

There are a fixed number of trials.
There are two possible outcomes (success or failure).
The trials are independent of each other.
There is a fixed probability of success.

1) A random variable $X \sim$ B(10, 0.6).
Find P$(X = 2)$ and P$(X > 8)$.

$$P(X = 2) = \binom{10}{2}(0.6)^2(0.2)^8 = \frac{81}{1\,953\,125}$$  $n = 10, p = 0.6$ and $r = 2$

$$P(X > 8) = P(X = 9) + P(X = 10)$$

$$\binom{10}{9}(0.6)^9(0.2)^1 + \binom{10}{10}(0.6)^{10} = 0.026 \ (3 \text{ s.f.})$$

2) A sample of 20 screws is checked for defects from a larger sample. It is claimed that only 2% will be defective. The sample is rejected if two or more screws are defective. Find the probability that the sample is rejected.

X is the number of screws out of 20 which are defective.

$X \sim$ B(20, 0.02)

$$P(X \geq 2) = 1 - P(X \leq 1) = 1 - P(X = 0) - P(X = 1)$$

$$1 - \binom{20}{0}(0.98)^{20} - \binom{20}{1}(0.98)^{19}(0.02)$$

$$1 - 0.9401010215 = 0.060 \ (3 \text{ s.f.})$$

- A cumulative probability function is the sum of the individual probabilities up to and including the value given.
- For the binomial distribution, there are tables in the formula booklet which give P$(X \leq x)$ for various values of n and p.

A biased dice is designed so that the probability of landing on a six is 0.3. Rosalie rolls the dice 20 times. Find the probability that Rosalie rolls at least five sixes.

$X =$ Number of sixes rolled out of 20

$X \sim$ B(20, 0.3)

$$P(X \geq 5) = 1 - P(X \leq 4)$$

$$1 - P(X \leq 4) = 1 - 0.2375 = 0.7625$$

When you look up a value in the table, it gives the probability up to and **including** that value.

Key Point

The notation used to represent a binomial distribution is $X \sim$ B(n, p), where n is the number of trials and p is the probability of success.

Key Point

n and p are called the parameters of the binomial distribution.

Key Point

Note the link with pure mathematics where you learnt $\binom{n}{r} = \dfrac{n!}{r!(n-r)!}$

This can be calculated using the appropriate button on your calculator.

Key Point

Always remember to define your variable, e.g. $X =$ number of sixes rolled out of 20.

Quick Test

1. The random variable $Y \sim$ B(8, 0.2). Find P$(Y = 1)$.
2. Write down the conditions to be met for a random variable to be modelled by a binomial distribution.
3. The random variable $X \sim$ B(10, 0.4). Find P$(X \leq 6)$.

Key Words

probability distribution
binomial distribution

Hypothesis Testing

You must be able to:

- Understand the language and concept of a hypothesis test and how a sample is used to make inferences about a population
- Carry out a one-tailed and two-tailed test for the proportion of a binomial distribution, including finding the critical region and interpreting the results.

Hypothesis Tests and the Critical Region

- A statistical hypothesis is a statement about a population parameter. You test the statement by carrying out a hypothesis test.
- A hypothesis test involves assuming the **null hypothesis** is true, then considering how likely your observed value (**test statistic**) was based on this assumption. If the likelihood of observing your value for your test statistic was too small, you would reject H_0 and make the conclusion that your result is significant and therefore your assumption was probably wrong. The threshold which helps you to make this decision is called the **significance level**. The significance level is usually 10%, 5% or 1%.

The steps of a hypothesis test are:

1. Identify the population parameter (p for a binomial distribution).

2. State the null and **alternate hypothesis**.

3. Specify the significance level.

4. Compare the test statistic with the critical value or the probability of the test statistic occurring with the significance level.

5. Make a conclusion stating whether the test statistic is significant or not **and** a comment in context.

- The **critical region** is the region which, if your test statistic falls within, would result in you rejecting the null hypothesis. The critical region is based on the significance level.
- You need to be able to calculate the probability of your test statistic taking a particular value if your null hypothesis is true. This probability is then compared to your significance level.
- An alternative is to find the values, based on your null hypothesis, which would be the first to fall within the critical region and compare this with the value of your test statistic.
- You are always trying to decide if your test statistic falls inside or outside the critical region.

An observation is taken from a binomial distribution B(10, p). The observation is used to test H_0: $p = 0.45$, H_1: $p > 0.45$

Using a 5% significance level, find the critical region for this test.

Assume H_0 is true and B(10, 0.45).

$P(X \geqslant 7) = 1 - P(X \leqslant 6) = 1 - 0.8980 = 0.102 = 10.2\%$

$P(X \geqslant 8) = 1 - P(X \leqslant 7) = 1 - 0.9726 = 0.0274 = 2.74\%$

$P(X \geqslant 7) > 5\%$ but $P(X \geqslant 8) < 5\%$

Therefore 8 is the critical value and 8, 9 or 10 would be the critical region.

> ### Key Point
>
> The **null hypothesis**:
> H_0: $\theta = s$
>
> The **alternate hypothesis**:
> H_1: $\theta \neq s$ (two-tailed) or
> H_1: $\theta > s$ or H_1: $\theta < s$ (one-tailed)
>
> Where θ is the population parameter and s is a given number dependent on the test.

> ### Key Point
>
> In a statistical test, evidence comes from a sample. The data from the sample is summarised as a test statistic.
>
> Examples of test statistics are the mean of a sample and the probability of success based on a sample.

> ### Key Point
>
> The significance level represents the percentage chance of you incorrectly rejecting H_0.

 If your observed value was 8, 9 or 10, you would reject H_0. If your observed value was anything less than 8, there would be insufficient evidence to reject H_0.

One and Two-tailed Hypothesis Tests

1) The standard treatment for a skin condition has a success rate of 55%. A doctor has designed a new drug and carried out an experiment in which 12 out of 20 patients were cured. She claims the new drug has a higher success rate than the old drug. Test, at a 5% significance level, if 12 out of 20 patients is enough evidence to support the doctor's claim.

   X represents the number of patients out of 20 who are cured.

   $X \sim B(20, 0.55)$

   $H_0: p = 0.55$, $H_1: p > 0.55$

   Option 1: work out the probability of 12 or more patients being cured on the old drug and compare to the significance level.

   $P(X \geqslant 12) = 1 - P(X \leqslant 11) = 1 - 0.8692 = 0.1308$

   13.08% > 5% therefore insufficient evidence to reject H_0. The doctor's claim cannot be supported.

   Option 2: find the critical value and compare to the test statistic (in this case 12).

   $P(X \geqslant 13) = 1 - P(X \leqslant 12) = 1 - 0.9420 = 0.058$

   $P(X \geqslant 14) = 1 - P(X \leqslant 13) = 1 - 0.9786 = 0.0214$

   0.058 > 0.05 but 0.0214 < 0.05

   Critical value is 14, so critical region is $X \geqslant 14$. 12 < 14 so there is insufficient evidence to reject H_0. The claim cannot be supported.

2) A single observation y is taken from a binomial distribution B(30, p). The value 5 is obtained. Test, at a 5% significance level, the following hypotheses:

   $H_0: p = 0.4$, $H_1: p \neq 0.4$

   Let C_1 and C_2 be the critical values.

   For the lower critical value:

   $P(Y \leqslant C_1) < 0.025$

   $P(Y \leqslant 7) = 0.0435 > 0.025$

   $P(Y \leqslant 6) = 0.0172 < 0.025$, therefore $C_1 = 6$.

   For the upper critical value:

   $P(Y \geqslant C_2) < 0.025$

   $P(Y \geqslant 17) = 1 - P(Y \leqslant 16) = 1 - 0.9519 = 0.0481 > 0.025$

   $P(Y \geqslant 18) = 1 - P(Y \leqslant 17) = 1 - 0.9788 = 0.0212 < 0.025$, therefore $C_2 = 18$.

   5 < 6, therefore 5 is in the critical region. There is enough evidence to reject the null hypothesis; it is likely that the observed value did not come from the distribution B(30, 0.4).

> **Key Point**
>
> A one-tailed test has one critical region at either end depending on the alternate hypothesis.
>
> A two-tailed test has two critical regions and the significance level is split equally between them.

Note this is one-tailed as the doctor is claiming the drug has improved the success rate.

This example demonstrates that when taking a sample from a population, there will be variations between samples. It is likely, based on a cure rate of 55%, that in a sample of 20, 12 people or more are likely to be cured so this is not enough evidence to suggest the new drug is any better.

This is two-tailed as you are testing $p \neq 0.4$ rather than greater than or less than.

> **Key Point**
>
> Remember when using the cumulative tables, the probability given is up to and including the value you have looked up.

Quick Test

1. Find the critical region for this hypothesis test at a 5% significance level. X is the test statistic, $X \sim B(20, p)$. $H_0: p = 0.2$, $H_1: p > 0.2$
2. Find the critical region for this hypothesis test at a 5% significance level. X is the test statistic, $X \sim B(40, p)$. $H_0: p = 0.3$, $H_1: p \neq 0.3$

Key Words

null hypothesis
test statistic
significance level
alternate hypothesis

Populations and Samples

1 Perri wants to investigate the average rainfall in Leuchars. The extract below shows her sample taken from the large data set.

Date	LEUCHARS Daily Mean Temperature (0900-0900) (°C)	Daily Total Rainfall (0900-0900) (mm)	© Crown Copyright Met Office 2015 Daily Total Sunshine (0000-2400) (hrs)	Daily Mean Windspeed (0000-2400) (kn)
01/05/2015	3.8	tr	9.1	8
02/05/2015	4.6	7.2	5.7	13
03/05/2015	9.3	15.4	0	15
04/05/2015	11.0	3.8	8.6	10
05/05/2015	7.5	16	1.3	12
06/05/2015	9.4	1	2.1	7
07/05/2015	6.8	tr	7.2	11
08/05/2015	4.1	5.6	5	7
09/05/2015	9.2	2	5.2	8
10/05/2015	9.8	3	0.4	11
11/05/2015	12.7	0.6	9.9	18
12/05/2015	11.5	tr	8.5	20
13/05/2015	9.5	tr	9.6	10
14/05/2015	9.1	0	0	8
15/05/2015	7.4	0.6	6.4	11
16/05/2015	9.6	0.2	12.2	16
17/05/2015	9.9	0.8	9.9	16
18/05/2015	10.0	2.4	2.6	8
19/05/2015	8.7	0.6	7.4	9
20/05/2015	11.9	1	9.1	10

a) Calculate the mean rainfall at Leuchars for the first 20 days in May, 2015. [2]

b) Explain how you dealt with the data for 01/05/2015, 07/05/2015, 12/05/2015 and 13/05/2015. [1]

c) State with a reason whether or not this sample is random. [2]

d) Suggest an alternative sampling method and give one advantage and one disadvantage of this technique in context. [3]

e) i) Perri uses her answer to part a) to estimate the average rainfall in Leuchars in the summer of 2015. Comment on the reliability of this estimate. [2]

   ii) How might this estimate be improved? [1]

Total Marks .... / 11

Measures of Central Tendency and Spread

1 Given that for a variable x, $\sum x = 21.4$, $\sum x^2 = 520$ and $n = 10$:

a) Find the mean. [2]

b) Find the variance. [3]

c) Find the standard deviation. [1]

2 Here is a set of data: 43   41   67   63   94   86   39

a) Calculate the mean. [2]

b) Find the median. [2]

3 The table summarises the number of breakdowns on a particular stretch of road over a period of time.

Number of breakdowns	0	1	2	3	4	5
Number of days	3	5	4	3	3	2

a) Calculate the mean. [3]

b) Calculate the standard deviation. [3]

4 The table summarises the distances thrown by the javelin throwers at Kingstown Athletics Club during a series of competitions.

a) Estimate the value of the median. [3]

b) Estimate the value of the interquartile range. [3]

Distance thrown (m)	Frequency
$15 \leqslant d < 20$	5
$20 \leqslant d < 25$	10
$25 \leqslant d < 30$	15
$30 \leqslant d < 35$	12
$35 \leqslant d < 45$	6

Total Marks _____ / 22

Displaying and Interpreting Data 1 & 2

1 A farmer weighed a random sample of young female pigs. The results are displayed in the histogram and partially summarised in the table.

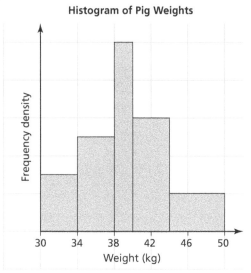

Histogram of Pig Weights

Weight (kg)	Frequency	Cumulative frequency
$30 \leqslant w < 34$		
$34 \leqslant w < 38$		
$38 \leqslant w < 40$	30	
$40 \leqslant w < 44$		
$44 \leqslant w < 50$		

a) Justify why a histogram is an appropriate diagram for this data. [1]

b) Complete the table. [3]

c) Estimate the percentage of pigs which weigh between 37 kg and 41 kg. [2]

d) On a separate sheet of paper, draw a cumulative frequency curve to represent this data. [2]

e) Use your curve to estimate the values for the lower quartile, median and upper quartile. [2]

2 In the sample in question 1, the smallest female pig weighed 30 kg and the largest female pig weighed 49 kg.

a) Given there are no outliers, draw a box and whisker plot to represent this data. [3]

Here is a box and whisker plot for the young male pigs:

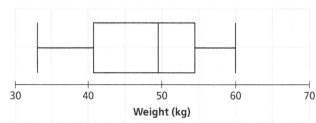

Weight (kg)

b) Compare the weights of the young female pigs with the weights of the young male pigs. [2]

Total Marks .... / 15

Calculating Probabilities

1. There are 30 pupils in a class and the languages spoken by each pupil were recorded and represented in a Venn diagram. The pupils spoke either French, Spanish, German or both French and Spanish.

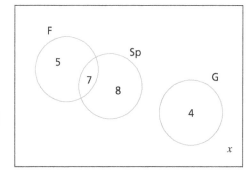

   a) Find the value of x. [2]

   b) A pupil is selected at random. Write down the probability that the pupil speaks both Spanish and German. [1]

   c) A pupil is selected at random. Find the probability that the pupil speaks French or Spanish. [2]

2. A box of chocolates contains 8 dark and 12 milk chocolates. Esmai selects two chocolates from the box.

   a) Draw a tree diagram on a separate sheet of paper to represent this information. [3]

   b) Work out the probability that she selects one of each type of chocolate. [3]

3. $P(A) = 0.5$ and $P(A \text{ and } B) = 0.3$.

   a) Given that events A and B are independent, find $P(B)$. [2]

   b) Draw a Venn diagram to represent these two events. [3]

   c) Find $P(A \text{ or } B \text{ or both})$. [2]

Total Marks _____ / 18

Statistical Distributions

1 The random variable $X \sim B(10, 0.1)$.

   a) Find $P(X = 3)$. [2]

   b) Find $P(X \leqslant 2)$. [1]

   c) Find $P(X \geqslant 6)$. [2]

2 Seven fair coins are tossed and the number of tails shown is recorded.

   Find the probability of:

   a) no tails [1]

   b) more than five tails [2]

   c) fewer than two tails. [2]

3 The random variable $Y \sim B(25, 0.2)$.

   a) Find the smallest value of r such that $P(Y > r) < 0.05$. [1]

   b) Find the greatest value of r such that $P(Y \leqslant r) < 0.05$. [1]

4 The random variable Q has the probability function

$$P(Q = q) = \begin{cases} \dfrac{2}{q}, & q = 4, 6, 12 \\ 0, & \text{otherwise} \end{cases}$$

   Construct a probability distribution table for Q. [2]

Total Marks _____ / 14

Hypothesis Testing

1. Explain what is meant by:

   a) a hypothesis test [1]

   b) a critical region [1]

   c) the significance level. [1]

2. A company claims that, on average, only 10% of seeds purchased from them will not germinate. Thomas purchases a packet of 50 seeds and 9 of them do not germinate. He complains to the company.

   Test, at the 5% significance level, if there is enough evidence to support his complaint. [5]

3. A single observation X is to be taken from a binomial distribution B(30, p).

   This observation is used to test: H_0: $p = 0.35$ H_1: $p \neq 0.35$

   a) Using a 5% significance level, find the critical region for this test. The significance level should be as close to 5% as possible. [3]

   b) State the actual significance level of this test. [1]

   The actual value of X obtained is 3.

   c) State a conclusion which can be drawn based on this value. [1]

   d) Give an interpretation of the actual significance level in this case. [1]

4. A marketing company claims that 'Crumbly Crumb' biscuits taste better than 'Chocolaty Choc' biscuits. Fifteen people were asked which they prefer and 10 said 'Crumbly Crumb'.

   Test, at a 5% significance level, if the marketing company's claim is true. State your hypothesis clearly. [7]

   Total Marks .... / 21

Indefinite Integrals

1 Evaluate:

a) $\int (x^6 - 2x + 5)\,dx$ [1]

b) $\int \dfrac{1}{3p^6}\,dp$ [1]

c) $\int \left(3t - \dfrac{1}{t^3} + 3\right)dt$ [1]

d) $\int \dfrac{1 - \sqrt{x}}{x^4}\,dx$ [1]

e) $\int (4s^{-\frac{1}{2}} + 5s)\,ds$ [1]

2 Given that $\dfrac{dy}{dx} = 5x^4 - \dfrac{3}{2}x^3 + \dfrac{4}{x^3}$, find y. [3]

3 Evaluate $\int \dfrac{6 - 2\sqrt{x}}{5x^2}\,dx$. [4]

4 **a)** Show that $\left(1 + 5\sqrt{x}\right)^2$ can be written in the form $1 + a\sqrt{x} + bx$, where a and b are constants to be found. [2]

b) Hence find $\int \left(1 + 5\sqrt{x}\right)^2 dx$. [3]

5 A curve passes through the point $(-2, 5)$ and the gradient function is $f'(x) = 4x - 2$. Find the equation of the curve. [3]

Total Marks _____ / 20

Integration to Find Areas

1 Calculate the following integrals:

a) $\displaystyle\int_{-1}^{3} (-3t^3 + 6)\,dt$ [2]

b) $\displaystyle\int_{10}^{20} (-q^2 + 5q)\,dq$ [2]

c) $\displaystyle\int_{1}^{3} \left(\dfrac{2}{x^3} - \dfrac{2}{7x^2}\right)dx$ [2]

d) $\displaystyle\int_{-1}^{2} \left(\dfrac{x^2 + x}{x^4}\right)dx$ [2]

2 Find the area of the finite region bounded by the curve with equation $y = 4x^2 + 3x - 1$, the x-axis and the lines $x = 4$ and $x = 1$. [4]

3 The diagram shows a sketch of the curve C with equation $y = x(x^2 - 9)$.

Use calculus to find the finite region bounded by the curve and the x-axis. Give your answer in the form $p + q\sqrt{x}$. [4]

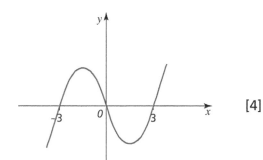

4 Find the area of the region bounded by the curve with equation $y = (3 - x)(x + 1)$, the positive x-axis and the positive y-axis. [4]

Total Marks .... / 20

Introduction to Vectors

1 Find a vector of magnitude 12 in the direction of **i** – **j**, giving your answer in rationalised surd form. [4]

2 Find the magnitude and direction of the vector **p** if:

a) $\mathbf{p} = \dfrac{\sqrt{3}}{2}\mathbf{i} + \dfrac{1}{2}\mathbf{j}$ [4]

b) $\mathbf{p} = 4\mathbf{i} - 4\mathbf{j}$ [4]

3 Find the magnitude and direction of the resultant of vectors **a** and **b**: [6]

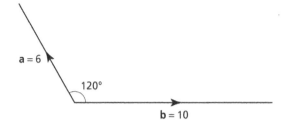

Total Marks _____ / 18

Vector Geometry

1 Given two points A($k + 1$, 1) and B($-k$, k), show that the shortest possible distance between

the two points is $\dfrac{3\sqrt{5}}{5}$. [7]

2 A parallelogram ABCD has coordinates A(−1, 4), B(3, 10) and D(8, −3).

Find the coordinates of C. [5]

3 A is the point (10, 9) and B is the point (8, 4).

Find the position vector of the point C given that B splits AC in the ratio 2 : 7. [5]

Total Marks _____ / 17

Proof

1 ABC is a right-angled triangle. Angle ACB = $x°$. Angle BAD = $90 - 2x°$. [3]

Prove that ACD is an isosceles triangle.

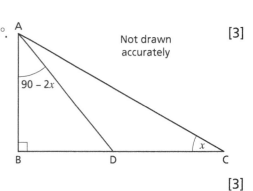

Not drawn accurately

$90 - 2x$

B   D   C

2 Show that $(2n - 1)^2 + (2n + 1)^2 = 8n^2 + 2$. [3]

3 Prove that, for any two numbers, the product of their difference and their sum is equal to the difference of their squares. [3]

4 Prove that, if the difference of two numbers is 4, then the difference of their squares is a multiple of 8. [3]

5 Sophie says that all square numbers are even. Use an example to prove that she is wrong. [1]

6 Explain the error made in this proof:

$$a = b$$
$$a^2 = ab$$
$$a^2 - b^2 = ab - b^2$$
$$(a - b)(a + b) = b(a - b)$$
$$a + b = b$$

As $a = b$

$$2b = b$$

If $b = 1$, this would suggest $2 = 1$ [2]

Total Marks _____ / 15

Quantities and Units in Mechanics

You must be able to:

- Understand and use quantities in the SI system: length, time and mass
- Understand and use the derived quantities of force and weight
- Understand and use position, displacement, distance travelled, velocity, speed and acceleration.

SI Units

- The SI system of units stands for *Système international* (French). It uses a metric system of units, all built upon seven base units.
- The three base units you will use in mechanics are length (the metre), mass (the kilogram) and time (the second). All other metric units used may be derived from these base units.

Force

- One form of Newton's second law of motion states that when a constant force is applied to a mass, it produces an acceleration proportional to the force applied, and inversely proportional to the mass.
- This can be expressed as $a \propto \frac{F}{m}$, or $F = kma$. 1 'Newton' is defined to be the force which, when acted on a 1kg mass, produces an acceleration of 1ms^{-2}.
- So inserting these values into $F = kma$, you have $1 = k \times 1 \times 1$. Therefore $k = 1$ and you obtain the formula $F = ma$.
- It is important to note that both acceleration and force are **vector quantities**, so when using the formula $F = ma$, their direction must be clearly indicated, either in a question or an answer or both. The force will be in the direction of the acceleration.

> **Key Point**
>
> It is also important to note that F stands for the resultant force on an object (i.e. all forces added together).

A mass of 5kg is pulled along a straight surface with force 5N. The drag force is a constant 2N. Find the acceleration of the mass.

$F = ma$ Using Newton's second law.

$F = 5 - 2 = 3N$

Therefore $3 = 5a$ and $a = \frac{3}{5}$ ms^{-2}.   The object will move to the right, and since F is the resultant force in this direction, you have 3N.

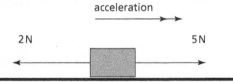

acceleration

2N   5N

Weight

- Weight is one type of force you use in mechanics. There are other types of force, such as tension, thrust and friction.
- The weight of a mass is the force on an object due to gravity.
- Applying $F = ma$ to a mass in the Earth's gravitational field, you obtain the formula $W = mg$, where W is the force (in Newtons), m is the mass (in kilograms) and g is the acceleration due to gravity, which is approximately 9.8ms^{-2}.
- So, for example, a mass of 1kg will weigh approximately $1 \times 9.8 = 9.8N$.
- On the Moon, the gravitational field strength is weaker, and the acceleration due to gravity is approximately 1.6ms^{-2}. Therefore, the same mass of 1kg would weigh approximately $1 \times 1.6 = 1.6N$.

> **Key Point**
>
> Although in mechanics we usually take g to be constant, in fact the value of g varies depending on where you are on the Earth. Near the Equator it is approximately 9.78ms^{-2}, whereas near the North Pole it is 9.83ms^{-2}. Therefore, you would feel 'heavier' the further north you travel.

Kinematics

- Kinematics refers to the science of movement.
- **Displacement** is a vector quantity. It tells you how far an object has been displaced from its original position.
- **Velocity** is a vector quantity. It tells you an object's speed in a given direction.
- Acceleration is a vector quantity. It tells you the object's acceleration in a given direction.

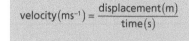

$$velocity(ms^{-1}) = \frac{displacement(m)}{time(s)}$$

$$acceleration(ms^{-2}) = \frac{velocity(ms^{-1})}{time(s)}$$

- The magnitude of the displacement tells you how far the object is from its initial position. The magnitude of the velocity is the object's speed.
- For example:

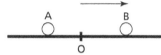

  - Ball A lies at a distance of 2 m from the centre O, while ball B's distance is 3 m from O.
  - Although the displacement of B is 3 m, the displacement of A is −2 m, as it is situated to the left of O.
- A similar example may be used as an illustration for velocity and speed:

> **Key Point**
>
> The value of *g* also decreases with altitude, although in any mechanics problems you encounter you should consider $g = 9.8\,ms^{-2}$, and assume *g* to be constant, unless stated otherwise.

Balls A and B (above) start moving from point O. Ball A moves with velocity −2 ms⁻¹, while ball B travels with speed 5 ms⁻¹.

After 10 seconds, find:

a) their displacements from O.

For ball A, displacement = −2 × 10 = −20 m

For ball B, displacement = 5 × 10 = 50 m

b) their distances from O.

For ball A, distance = 20 m (to the left of O)

For ball B, distance = 50 m (to the right of O)

c) how far apart they lie.

They lie a distance 20 + 50 = 70 m apart

> You can use the formula:
> displacement = velocity × time

> Since distance is the magnitude of displacement.

> **Quick Test**
>
> 1. An object has acceleration −0.5 ms⁻². Given that it starts from rest at a point O, find:
>   a) its velocity after 12 seconds
>   b) its speed after 10 seconds.
> 2. An object of mass 5 kg is at rest at point O. 20 seconds later it has a velocity of 25 ms⁻¹. Find the magnitude of the force acting on the object, assuming it is constant.
> 3. Falling objects on the planet Mars have an acceleration of 3.7 ms⁻². How much less would a 60 kg girl weigh on Mars, compared to Earth?

> **Key Words**
>
> force
> acceleration
> mass
> vector quantity
> displacement
> velocity

Displacement–time and Velocity–time Graphs

You must be able to:

- Understand and use displacement–time graphs
- Understand and use velocity–time graphs.

Displacement–time Graphs

- A **displacement–time graph** represents the motion of an object, with displacement on the y-axis, and time on the x-axis.
- The gradient of the graph tells you the velocity of the object at any particular moment.
- The diagram (right) illustrates the motion of a white van.
- The motion can be split into three separate sections (from O to A, A to C and C to D):
  - From O to A, the graph is a straight line, indicating a constant velocity. The velocity at which the van is moving is the gradient of the graph, i.e. $\frac{10}{8} = 1.25 \text{ ms}^{-1}$.
  - From A to C, the graph is again a straight line, but this time with negative gradient. This indicates the van is moving backwards, having negative velocity, i.e. $-\frac{18}{20} = -0.9 \text{ ms}^{-1}$. At point B the van returns to its original starting position.
  - From C to D, the graph is a horizontal line, indicating the van remains stationary for 12 seconds.

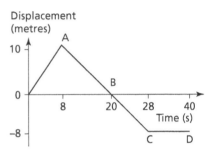

Velocity–time Graphs

- A **velocity–time graph** represents the motion of an object, with velocity on the y-axis, and time on the x-axis.
- The **gradient** of the graph tells you the acceleration of the object at any particular moment.
- The **area** under the graph tells you the total distance travelled by the object.
- The diagram (right) illustrates the motion of a lift travelling vertically upwards:
  - O to A is a straight line, indicating constant acceleration.
  - The acceleration is the gradient of the line, i.e. $\frac{1.5}{3} = 0.5 \text{ ms}^{-2}$.
  - The line from A to B is horizontal, indicating the lift is moving with a constant velocity of 1.5 ms^{-1}.
  - From B to C the line has a negative gradient, indicating an acceleration of $-\frac{1.5}{4} = -0.375 \text{ ms}^{-2}$.

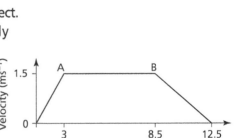

  - The total distance travelled by the lift is equal to the area under the curve. Treating this graph as a trapezium, you have the area:

  $A = \frac{1}{2} \times (5.5 + 12.5) \times 1.5 = 13.5$. So this represents a distance of 13.5 m.

- The **average speed** of the lift is equal to $\frac{\text{total distance}}{\text{total time}} = \frac{13.5}{12.5} = 1.08 \text{ ms}^{-1}$.

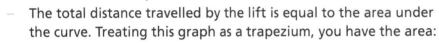

Two Moving Objects

- Very often you will be asked to illustrate the motion of two objects on the same velocity–time graph, and use it to answer various questions.
- When doing so, it is important to mark clearly on the axes the values of velocity and time relevant to each object.

At time $t = 0$, cyclist Bill, travelling at a constant speed of $6\,\text{ms}^{-1}$, overtakes cyclist Beth, who is travelling at a constant speed of $4.5\,\text{ms}^{-1}$. After two minutes, Beth decelerates at $0.3\,\text{ms}^{-2}$, before coming to rest.

In order to reach Beth, Bill cycles for T seconds, before decelerating at a constant rate and coming to rest at the same time as Beth.

a) On the same axes, draw velocity–time graphs showing the motion of both Bill and Beth.

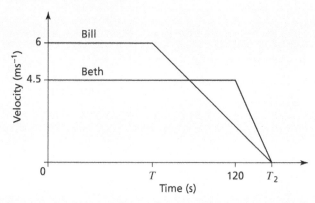

b) Find the time at which Beth comes to rest.

$$0.3 = \frac{4.5}{t} \text{ and } t = \frac{4.5}{0.3} = 15 \text{ seconds}$$

Therefore $T_2 = 120 + 15 = 135$ seconds

 To find T_2, use deceleration
$$= \frac{\text{decrease in velocity}}{\text{time}}$$

c) Find the value of T.

$$6T + \frac{1}{2} \times 6 \times (135 - T) = \frac{1}{2} \times (120 + 135) \times 4.5$$

 To find T, use the fact that both Bill and Beth have travelled the same distances. So the areas under each graph must be equal.

$$6T + 405 - 3T = 573.75$$
$$3T = 168.75$$
$$T = 56.25 \text{ seconds}$$

Key Point

Remember that the average speed of motion is $\dfrac{\text{total distance}}{\text{total time}}$.
It is not equal to the 'average of the speeds' from each part of the motion.

Key Point

The fact that the acceleration is negative in a velocity–time graph does not mean the object is travelling backwards. It is 'slowing down', or decelerating.

Key Point

With questions such as part **a)** (above left), it is tempting to draw the graphs like this:

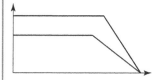

However, this cannot be correct since the distance travelled by each cyclist must be equal, and the areas under these graphs cannot be equal.

Quick Test

1. Draw an **acceleration–time graph** for the vertical lift example described on page 112.
2. A walker starts moving from rest with constant acceleration for T seconds, reaching a speed of $3\,\text{ms}^{-1}$. He maintains this speed for another $2T$ seconds, walking a total distance of $100\,\text{m}$.
   a) Sketch a velocity–time graph for the walker.
   b) Calculate the value of T.

Key Words

displacement–time graph
velocity–time graph
gradient
area
average speed

Kinematics with Constant and Variable Acceleration

You must be able to:

- Derive and use the constant acceleration equations for motion in a straight line
- Use calculus in kinematics for motion in a straight line.

Constant Acceleration Equations

- You need to be able to derive each of the five constant acceleration equations, commonly referred to as *suvat* equations. Each letter stands for a property of motion as shown in the table.
- Each equation may be derived from a general velocity–time graph.
- Consider an object moving with constant acceleration a, from an initial velocity u through to a final velocity v. Suppose the motion takes a total time t. The velocity–time diagram looks as below right:

Letter	Quantity	SI units
s	displacement	m
u	initial velocity	ms⁻¹
v	final velocity	ms⁻¹
a	acceleration	ms⁻²
t	time	s

  – The gradient gives you the acceleration, so $a = \dfrac{v-u}{t}$, or:

1 $v = u + at$

  – The area under the graph gives you the displacement, so:

2 $s = \frac{1}{2}(u+v)t$

  – Now if you substitute equation 1 into equation 2:
  $s = \frac{1}{2}\big(u+(u+at)\big)t$, or $s = \frac{t}{2}(2u+at)$, giving:

3 $s = ut + \frac{1}{2}at^2$

  – Equation 1 may be rearranged to give $t = \dfrac{v-u}{a}$

  – Substituting this into equation 2 gives $s = \frac{1}{2}(u+v)\left(\dfrac{v-u}{a}\right)$
  $2as = (u+v)(v-u) \therefore 2as = v^2 - u^2$, so:

4 $v^2 = u^2 + 2as$

  – Finally, equation 1 gives you $u = v - at$. Substituting this into equation 3 will give you:

5 $s = vt - \frac{1}{2}at^2$

- Notice that each equation connects four of the five possible variables. Therefore, in any constant acceleration problem, if you are given three pieces of information, you can calculate each of the others.

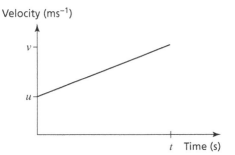

Velocity (ms⁻¹)

t Time (s)

> **Key Point**
>
> Practise deriving these equations yourself. This is a new part of the AS syllabus, so it has a good chance of being tested.

A car accelerates from $30\,\text{ms}^{-1}$ to $56\,\text{ms}^{-1}$ in 10 seconds. Find:

a) the distance travelled by the car

$s = \frac{1}{2}(u+v)t$

> You know $u = 30$, $v = 56$, $t = 10$ and wish to find s, so use $s = \frac{1}{2}(u+v)t$.

$s = \frac{1}{2}(30 + 56) \times 10 = 430\,\text{m}$

b) the acceleration of the car.

$v = u + at$

> You know $u = 30$, $v = 56$, $t = 10$ and wish to find a, so use $v = u + at$.

$a = \dfrac{v-u}{t} = \dfrac{56-30}{10} = 2.6\,\text{ms}^{-2}$

> **Key Point**
>
> In part **b)** in this example, you could have used any other equation. However, it is best to use the original information given in the question (if you can) in case you make an error in part a).

Calculus in Straight-line Kinematics

- For cases where an object has non-constant acceleration (or non-constant velocity), calculus may be used. In such cases, you normally denote r for the displacement, though s may also be seen.
- You can use the results $v = \dfrac{dr}{dt}$ and $a = \dfrac{dv}{dt}$

  Therefore, you also have the results $r = \int v\, dt\, (+c)$ and $v = \int a\, dt\, (+c)$.
- These results also hold for cases of constant acceleration, e.g. you may be expected to derive constant acceleration equations 1 and 3 using them.
- Suppose an object moves with constant acceleration a:
  - Then using $v = \int a\, dt\, (+c)$, you have $v = at + c$.
  - If at time $t = 0$, $v = u$, then $u = 0 + c$ (i.e. $v = u + at$, giving equation 1).
  - Writing v as $\dfrac{dr}{dt}$, you have $\dfrac{dr}{dt} = u + at$, so $r = \int (u + at)\, dt$, <---

    i.e. $r = ut + \frac{1}{2}at^2 + c$.

    > Remember, when integrating, that in this case both u and a are constants.

  - If at time $t = 0$, the displacement $= 0$, then $c = 0$ and you have $r = ut + \frac{1}{2}at^2$, which is equation 3.

1) A particle moves in a straight line such that its displacement from a point O after a time t seconds is given by $s = 5 + 4t^2 - 2t^3$ for $0 \leqslant t < 5$.

   > An example using non-constant acceleration.

   a) Find the time(s) when the particle is momentarily stationary.

   When the particle is stationary, $v = 0$, and so $\dfrac{ds}{dt} = 0$.

   $\dfrac{ds}{dt} = 8t - 6t^2$

   $8t - 6t^2 = 0 \therefore 2t(4 - 3t) = 0$ <--- Factorise in order to solve for t.

   So $t = 0$ or $t = \frac{4}{3}$

   b) Find the particle's acceleration at time $t = \frac{2}{3}$ and interpret your result.

   $a = \dfrac{dv}{dt} = 8 - 12t$ <--- To find the acceleration, you differentiate again.

   At time $t = \dfrac{2}{3}$, $\dfrac{dv}{dt} = 8 - 12 \times \left(\dfrac{2}{3}\right) = 0$

   Therefore at this time, the acceleration is 0, so in this case the particle has reached its maximum (or minimum) velocity.

2) A particle is moving in a straight line with velocity $v = 2t^3 - 3t$. At time $t = 1\,s$, the displacement $r = 5\,m$. Find r at time $t = 2$.

   > An example involving integration.

   $r = \int v\, dt = \int (2t^3 - 3t)\, dt = \dfrac{t^4}{2} - \dfrac{3t^2}{2} + c$ <---

   > Always remember to include a constant of integration.

   Substituting $t = 1$, $r = 5$ gives $5 = \dfrac{1}{2} - \dfrac{3}{2} + c$, so $c = 6 \therefore r = \dfrac{t^4}{2} - \dfrac{3t^2}{2} + 6$

   At time $t = 2$, $r = \dfrac{2^4}{2} - \dfrac{3 \times 2^2}{2} + 6 = 8\,m$.

Quick Test

1. A particle travels in a straight line with a constant deceleration of $3\,ms^{-2}$. Given that its initial velocity is $10\,ms^{-1}$, find the first time when its displacement from its starting point is $4\,m$.
2. A particle travels in a straight line from point O with variable acceleration $a = 3t\,ms^{-2}$. Given that initially it has a velocity of $-2\,ms^{-1}$, find the time when it is momentarily stationary.

Key Words

constant acceleration

Forces and Newton's Laws 1

You must be able to:

- Apply Newton's laws of motion to a variety of situations
- Apply Newton's second law to motion in a straight line
- Apply constant acceleration equations to problems involving vertical motion and weight.

Newton's Laws of Motion

- Newton's first law states that an object will remain at rest or move with constant velocity, unless acted upon by a **resultant force**.
- Newton's second law states that for a constant resultant force, F:
  $F = ma$, where a is the **acceleration** in the direction of the resultant force.
- Newton's third law states that for every action, there is an equal and opposite reaction.
- Consider a child of weight 50 N standing on the ground:
  - The child's weight exerts a force of 50 N vertically downwards. By Newton's third law, there must be an upwards force R, otherwise the child would simply fall into the ground.
  - Since the child is stationary, the resultant force on her must be zero. Therefore $R = 50$ N.
- Now consider a man, this time standing in a lift, where the lift has acceleration 1.5 ms^{-2} upwards. Suppose the mass of the man is 75 kg:
  - The man's weight (downwards) will therefore be $75g$ N.
  - Since the man is accelerating, you apply Newton's second law, i.e. $F = ma$ in the direction of the acceleration (vertically upwards).
  - The resultant force upwards is $R - 75g$ N.
  - Therefore $R - 75g = 75 \times 1.5$ and so $R = 848$ N.

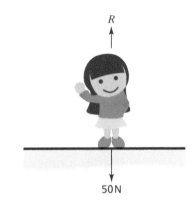

R

50 N

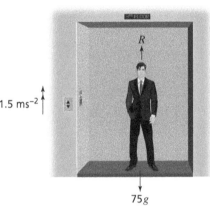

1.5 ms^{-2}

R

$75g$

Forces as Vectors

- Since force is a vector quantity, it is possible to write Newton's second law in vector form, i.e. $\mathbf{F} = m\mathbf{a}$.
- Either of $\mathbf{F}$ or $\mathbf{a}$ may be expressed as vectors, and you should be prepared to work with them both in this form.
- Likewise, you can use constant acceleration equations in vector form, with the exception of $v^2 = u^2 + 2as$, as it is not possible to 'square' a vector in this context.

1) A force of $12\mathbf{i} - 18\mathbf{j}$ acts on a 4 kg mass from rest. Find the distance the mass lies from its starting point after three seconds.

Applying $\mathbf{F} = m\mathbf{a}$, you have $12\mathbf{i} - 18\mathbf{j}$, so $\mathbf{a} = \dfrac{12\mathbf{i} - 18\mathbf{j}}{4} = 3\mathbf{i} - \dfrac{9}{2}\mathbf{j}$

Now using $\mathbf{s} = \mathbf{u}t + \dfrac{1}{2}\mathbf{a}t^2$ with $\mathbf{u} = 0$:

$\mathbf{s} = \dfrac{1}{2}\left(3\mathbf{i} - \dfrac{9}{2}\mathbf{j}\right) \times 9 = \dfrac{27}{2}\mathbf{i} - \dfrac{81}{4}\mathbf{j}$

This is the displacement vector, so now use Pythagoras' theorem to find the distance.

Therefore the distance from its starting point is

$|\mathbf{s}| = \sqrt{\left(\dfrac{27}{2}\right)^2 + \left(\dfrac{81}{4}\right)^2} = 24.3$ m

Key Point

It is convention to denote acceleration by a 'double arrow' on mechanics diagrams. This also indicates the direction in which you must calculate the 'resultant force'.

Key Point

Unless stated otherwise, always take $g = 9.8$ ms^{-2} and give your final answer to 3 s.f.

2) A particle of mass 2.5 kg has its velocity changed from $(-7\mathbf{i} - 3\mathbf{j})$ ms⁻¹ to $(20\mathbf{i} - 8\mathbf{j})$ ms⁻¹ in 5 seconds by a constant force, **F**.

a) Find the magnitude of the force (assumed to be constant).

Using $\mathbf{a} = \dfrac{\mathbf{v} - \mathbf{u}}{t}$

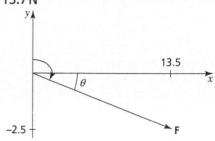

You may find it clearer and easier to work through problems such as these by using column vector notation. For example, a velocity of $(20\mathbf{i} - 8\mathbf{j})$ ms⁻¹ can be written as $\begin{pmatrix} 20 \\ -8 \end{pmatrix}$ ms⁻¹.

$$\mathbf{a} = \frac{1}{5}\left[\begin{pmatrix} 20 \\ -8 \end{pmatrix} - \begin{pmatrix} -7 \\ -3 \end{pmatrix}\right] = \frac{1}{5}\begin{pmatrix} 27 \\ -5 \end{pmatrix} = \frac{27}{5}\mathbf{i} - \mathbf{j}$$

Therefore $\mathbf{F} = m\mathbf{a} = \dfrac{5}{2}\begin{pmatrix} \frac{27}{5} \\ -1 \end{pmatrix} = \begin{pmatrix} 13.5 \\ -2.5 \end{pmatrix} = 13.5\mathbf{i} - 2.5\mathbf{j}$ N

The magnitude of the force will therefore be $|\mathbf{F}| = \sqrt{13.5^2 + 2.5^2} = 13.7$ N

b) Find the bearing in which the force acts.

To find the bearing, draw a diagram for **F** and apply right-angled trigonometry.

$\mathbf{F} = 13.5\mathbf{i} - 2.5\mathbf{j}$

Bearing $= 90 + \tan^{-1}\left(\dfrac{2.5}{13.5}\right) = 100°$ (3 s.f.)

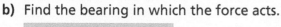

Forces in Equilibrium

- A set of forces is said to be in **equilibrium** if the algebraic sum of the forces is zero. This is a consequence of Newton's first law of motion. It effectively means that the sum of **i** components, and the sum of **j** components, must both be equal to zero.

Suppose forces $\mathbf{F}_1$, $\mathbf{F}_2$ and $\mathbf{F}_3$ act on a particle:

$\mathbf{F}_1 = (4\mathbf{i} + 5\mathbf{j})$ N, $\mathbf{F}_2 = (-2\mathbf{i} + 9\mathbf{j})$ N and $\mathbf{F}_3 = (a\mathbf{i} + b\mathbf{j})$ N

Given that the particle is in equilibrium, find the values of a and b.

The particle is in equilibrium, so $\mathbf{F}_1 + \mathbf{F}_2 + \mathbf{F}_3 = 0$

$\Rightarrow (4\mathbf{i} + 5\mathbf{j}) + (-2\mathbf{i} + 9\mathbf{j}) + (a\mathbf{i} + b\mathbf{j}) = 0$

$\Rightarrow 4\mathbf{i} - 2\mathbf{i} + a\mathbf{i} + 5\mathbf{j} + 9\mathbf{j} + b\mathbf{j} = 0$

$\Rightarrow (4 - 2 + a)\mathbf{i} + (5 + 9 + b)\mathbf{j} = 0$

$\Rightarrow (2 + a)\mathbf{i} + (14 + b)\mathbf{j} = 0$

So $2 + a = 0 \Rightarrow a = -2$ and $14 + b = 0 \Rightarrow b = -14$.

Alternative method using column vectors:

$\begin{pmatrix} 4 \\ 5 \end{pmatrix} + \begin{pmatrix} -2 \\ 9 \end{pmatrix} + \begin{pmatrix} a \\ b \end{pmatrix} = \begin{pmatrix} 0 \\ 0 \end{pmatrix}$

$\Rightarrow 4 - 2 + a = 0$

$\Rightarrow 5 + 9 + b = 0$

So $a = -2$ and $b = -14$.

Quick Test

1. A particle has constant acceleration $\begin{pmatrix} 6 \\ 2 \end{pmatrix}$ ms⁻². Given that its final velocity after two seconds is $\begin{pmatrix} 10 \\ 12 \end{pmatrix}$ ms⁻¹, find the particle's distance from its starting point at this time.

2. Forces $\mathbf{F}_1 = 2a\mathbf{i} + 5b\mathbf{j}$ and $\mathbf{F}_2 = 7b\mathbf{i} + (a + 3)\mathbf{j}$ act on a particle that remains in equilibrium. Find the values of a and b.

3. A car is driven in a straight line, accelerating at a rate of 0.5 ms⁻². Given that the car has a mass of 0.5 tonnes and encounters resistive forces totalling 150 N, calculate the driving force of the car.

Key Words

resultant force
equilibrium

Forces and Newton's Laws 2

You must be able to:

- Solve problems involving vertical motion under gravity
- Understand the motion of connected particles.

Vertical Motion Under Gravity

- For problems involving motion under **gravity**, you assume that the acceleration due to gravity is constant and therefore constant acceleration equations may be used.

A particle is projected vertically upwards with initial speed $50\,\text{ms}^{-1}$.

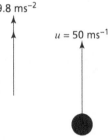

−9.8 ms⁻²

$u = 50\ \text{ms}^{-1}$

a) Find the **maximum height** reached by the particle.

Using $v^2 = u^2 + 2as$

$0 = 50^2 + 2 \times (-9.8) \times s$

$19.6s = 2500$

$\Rightarrow s = 128$ m

> At the maximum height, the particle's velocity is instantaneously zero. You therefore know u, v and a, and wish to find s.

b) Find the time taken for it to return to its starting position.

So $0 = 50t + \dfrac{1}{2}(-9.8)t^2$

Factorising gives $0 = t\,(50 - 4.9t)$

Therefore, $t = 0$ or $t = \dfrac{50}{4.9} = 10.2\,\text{s}$

To find the time when it returns to its starting point, use $s = ut + \dfrac{1}{2}at^2$
Remember, s stands for displacement (not distance travelled in this case), so $s = 0$.

> The first value gives the time before the particle has even moved, so you can reject this solution. The required answer is therefore 10.2 s.

- Vertical motion equations may lead to quadratic equations, which can be solved using the quadratic formula.

A particle is projected vertically upwards at a speed of $50\,\text{ms}^{-1}$. Find for how long its height is at least 100 m above the ground.

Using $s = ut + \dfrac{1}{2}at^2$, you obtain $100 = 50t - 4.9t^2$.

Rearranging this, you obtain the quadratic equation $4.9t^2 - 50t + 100 = 0$.

So $t = \dfrac{-b \pm \sqrt{b^2 - 4ac}}{2a} = \dfrac{-(-50) \pm \sqrt{(-50)^2 - 4 \times 4.9 \times 100}}{2 \times 4.9}$

This leads to two values: $t_1 = 2.7308$ and $t_2 = 7.4733$

Therefore, the length of time the particle lies above 100 m is equal to

$t_2 - t_1 = 7.4733 - 2.7308 = 4.74$ seconds

> You first find the time(s) when the particle is exactly 100 m above the ground.

Connected Particles

- **Connected particle** questions will involve two particles that are connected by light string (has no mass), so the weight of the string does not need to be taken into account.
- There are three 'standard' situations which you must be familiar with, though these are by no means exhaustive. In each case, the technique to use would be to draw a clear force diagram, then apply Newton's second law to each particle separately. It is likely you will then need to solve these equations simultaneously in order to find various required values.

Key Point

Remember, if the particle is falling freely under gravity, its acceleration is $9.8\,\text{ms}^{-2}$, but if it is being projected vertically upwards, its acceleration is $-9.8\,\text{ms}^{-2}$.

1. Connected Particles Travelling Horizontally

A 4kg mass and a 6kg mass are joined by a piece of string. The 4kg mass is pulled with a 20N force in a direction along a straight line, away from the 6kg mass. Each particle encounters resistances of 5N. By modelling the masses as particles and the string as being light, determine the acceleration of the system and the tension in the string.

The string is uniform throughout, so the tension is the same throughout.

First draw a complete force diagram with the information given in the question.

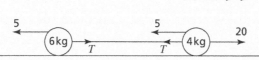

Applying $F = ma$ to the 4kg mass: $20 - T - 5 = 4a$ **(1)**

Applying $F = ma$ to the 6kg mass: $T - 5 = 6a$ **(2)**

Remember, F is the resultant force in the direction of the acceleration, so $F = 20 - T - 5$.

Now adding both equations gives: $20 - 10 = 10a$, so $a = 1\,\text{ms}^{-2}$.

Substituting this into equation **(2)** gives: $T - 5 = 6 \times 1$, so $T = 11\,\text{N}$.

2. Connected Particles Travelling Vertically (Pulley System)

- These situations involve two masses connected by a light string, hanging over a smooth pulley. Consider two particles, of masses 6kg and 2kg respectively, in this situation. The particles are held at the same level and then released. As in the previous example, you can use this initial information to determine the tension in the string and the acceleration of the system:

  – First draw a force diagram (see right).

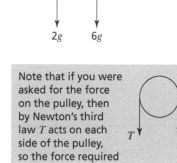

  – Now apply $F = ma$ to the 6kg mass: $6g - T = 6a$ **(1)**

  – Applying $F = ma$ to the 2kg mass: $T - 2g = 2a$ **(2)**

  – Now adding both equations gives: $6g - 2g = 8a$, so $a = \dfrac{g}{2} = 4.9\,\text{ms}^{-2}$.

  – Substituting into equation **(2)** gives: $T - 2g = 2\left(\dfrac{g}{2}\right)$, so $T = 3g = 29.4\,\text{N}$.

Note that if you were asked for the force on the pulley, then by Newton's third law T acts on each side of the pulley, so the force required would be $2T = 6g\,\text{N}$.

3. Connected Particles Travelling Both Horizontally and Vertically

The diagram shows two particles, P and Q, with masses of 4kg and 3kg respectively. The system is released from rest.

Find the acceleration of each particle.

R is not acting in the direction of the acceleration, so it is not part of the resultant force here.

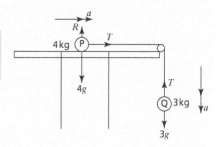

Applying Newton's 2nd law to P: $T = 4a$ **(1)**

Applying Newton's 2nd law to Q: $3g - T = 3a$ **(2)**

Adding these equations: $3g = 7a$, so $a = \dfrac{3g}{7}\,\text{ms}^{-2}$

Quick Test

1. A particle is dropped from the top of a tower, 158m high. How long will it take to hit the ground, and what will its velocity be when it hits the ground? What assumptions are you making to answer this?

2. In example 2 above, what should the ratio of the masses be if an acceleration of $\dfrac{g}{3}\,\text{ms}^{-2}$ is required?

Key Words

gravity
maximum height
connected particles
pulley

Quantities and Units in Mechanics

1 A mass of 10 kg is pulled along some rough ground in a straight line, with force 12 N.

a) Given that the mass moves with constant velocity, calculate the resistive force on the object.   **[2]**

b) Suppose the resistive force remains constant and the pulling force increases to 15 N.

Calculate the acceleration of the object.   **[3]**

2 A mass travels in a straight line with velocity $-3.5\,ms^{-1}$. Given that it starts from a point on the x-axis where $x = 3$, find the time taken by the mass to travel to the point $x = -12$.   **[3]**

> **Total Marks** _____ / 8

Displacement–time and Velocity–time Graphs

1 A car travels on a straight horizontal road. Its initial speed is $10\,ms^{-1}$, and it accelerates at a constant rate for 5 seconds, reaching a speed of $V\,ms^{-1}$. It maintains this speed for a further 10 seconds, before decelerating for a further 5 seconds at a constant rate until stationary. The car travels a total distance of 850 m.

a) Draw a velocity–time diagram for the car.   **[3]**

b) Find the value of V.   **[5]**

c) Find the car's final deceleration.   **[2]**

2 Trains A and B move along straight parallel tracks. Initially, train A travels at a speed of $50\,ms^{-1}$ and B is at rest. A is next to B on the platform. B then accelerates at a constant rate for T seconds until reaching a speed of $60\,ms^{-1}$, which it maintains until it catches up with train A.

After three minutes in total, both trains are again adjacent to each other.

a) Draw a velocity–time graph to illustrate the journeys of both trains on the same axes. [3]

b) Find time T. [5]

c) Find the time at which both trains have the same speed. [2]

Total Marks .... / 20

Kinematics with Constant and Variable Acceleration

1 A particle moves with a constant acceleration of $1.5\,ms^{-2}$.

If initially its speed is $2\,ms^{-1}$, find:

a) the distance the particle has travelled after 10 seconds [2]

b) the velocity of the particle when it has travelled half its total distance. [2]

2 A particle moves in a straight line such that its velocity at time t is given by $v = 2t^2 - 9t + 4$. Initially its displacement from a fixed point, O, is $\frac{5}{2}$ m.

Find:

a) the times when the particle is stationary [3]

b) its position after two seconds. [5]

Total Marks .... / 12

Forces and Newton's Laws 1

1 A force of magnitude 10 N changes the velocity of a particle of mass m kg from $(2\mathbf{i} + \mathbf{j})\,\text{ms}^{-1}$ to $(10\mathbf{i} + 6\mathbf{j})\,\text{ms}^{-1}$ in two seconds.

Find the value of m. [5]

2 A 3 kg mass (A) is placed on the floor of a lift, and a 2 kg mass (B) is placed on the 3 kg mass. The lift accelerates vertically upwards at a rate of $0.5\,\text{ms}^{-2}$.

Find:

a) the reaction force exerted on B by A [3]

b) the reaction force exerted on A by the floor of the lift. [3]

Total Marks .... / 11

Forces and Newton's Laws 2

1 A ball of mass $2m$ kg and a ball of mass m kg are joined by a piece of string. The $2m$ kg mass is pulled with a 10 N force in a direction along a straight line, in a direction away from both masses.

Given that the force produces an acceleration of 0.5 ms⁻², find:

a) the value of m [5]

b) the tension in the string. [1]

2 Joey and Alexia stand on top of a high building. Joey drops a small ball from rest. Five seconds later, Alexia throws an identical ball downwards with a speed of 2 ms⁻¹.

Given that both balls hit the ground at the same time, determine the height of the building. (Assume $g = 10$ ms⁻²). [7]

Total Marks .... / 13

Populations and Samples

1 Briefly explain what you understand by:

    a) a population [1]

    b) a sample. [1]

2 The 280 members of a badminton club are listed alphabetically in the club's membership booklet. The committee wishes to select 40 members to survey them about the facilities in the club.

    a) Explain how the committee could select a simple random sample of 40 members. [3]

The chairman argues that this sample might not fully represent all aspects of the population.

He presents the table here and suggests using a stratified sampling method.

	Male	Female
Under 30	80	60
Over 30	90	50

    b) Using the table, explain how the committee could take a stratified sample of 40 members. [4]

3 Give an example of a non-random sampling technique and an example in which it would be appropriately used. [2]

4 The table below shows the daily mean air temperature in Beijing for the first 24 days in May, 1987, from the large data set.

Day	1	2	3	4	5	6	7	8	9	10	11	12
Daily mean air temperature (°C)	13.7	12.7	17.0	17.0	19.8	21.0	21.5	21.6	21.2	20.7	21.6	22.1
Day	13	14	15	16	17	18	19	20	21	22	23	24
Daily mean air temperature (°C)	17.6	20.3	21.3	15.6	17.6	15.8	22.2	22.1	23.8	19.0	21.2	19.3

© Crown Copyright, Met Office 2015

    a) Use an opportunity sample of the first six days in the table to estimate the mean daily mean air temperature. [2]

b) Calculate the mean of all 24 days given to estimate the daily mean air temperature.

Comment on the reliability of your two estimates of the mean. [3]

Total Marks .... / 16

Measures of Central Tendency and Spread

1 Isobel records the time spent in the lunch hall by the students in her year. This table summarises her results:

Time (minutes)	20	21	22	23	24	25
Frequency	3	17	29	34	26	10

a) Calculate the mean time spent in the lunch hall. [2]

b) Calculate the standard deviation of the time spent in the lunch hall. [3]

2 A frequency distribution is shown here:

Class interval	1–10	11–20	21–30	31–40	41–50
Frequency	11	15	20	19	6

a) Estimate the 40th percentile. [2]

b) Estimate the 80th percentile. [2]

c) Estimate the 40th to 80th percentile. [1]

3 The times from a cross-country race are shown in the table.

Time taken (minutes)	Frequency
20–29	8
30–39	12
40–49	35
50–59	18
60–69	9

a) Estimate the mean time taken. [2]

b) Estimate the standard deviation. [3]

c) Estimate the number of runners whose race time is more than one standard deviation greater than the mean. [3]

Total Marks .... / 18

Displaying and Interpreting Data 1 & 2

1 Data for the daily mean temperature from Camborne in May, 1987, is taken from the large data set.

10.7	8.9	8.1	8.2	9.8	9.3	(Temperatures in °C)
10.9	10.5	10.9	9.9	8.8	10.2	
9.2	10.2	9.6	8.7	9.7	10.4	
9.5	11.1	10.5	11.1	9.8	10.2	
12.6	10.4	11.3	12.1	12.0	11.6	© Crown Copyright, Met Office 2015

a) Calculate the lower quartile and upper quartile. [3]

An outlier is defined as a value which lies either 1.5 × interquartile range above the upper quartile or 1.5 × interquartile range below the lower quartile.

b) Determine if the data contains any outliers. [2]

c) Draw a box and whisker diagram to represent this data. [3]

2 This data shows daily maximum gust plotted against daily mean wind speed for Heathrow in July, 1987.

Daily maximum gust (kn)	16	14	16	16	20	10	17	14	17	23	28	15	17	23	25	20	10	17	24	31	19	20	15	13	16	21	28
Daily mean wind speed (Beaufort conversion)	5	5	5	4	6	3	6	5	5	7	8	4	5	8	8	3	3	4	8	9	7	8	5	4	5	7	9

© Crown Copyright, Met Office 2015

The data has been plotted in this scatter diagram.

a) Describe the correlation between daily maximum gust and daily mean wind speed. [1]

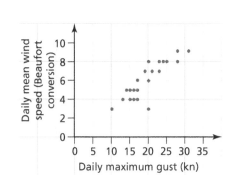

The line of regression has equation $w = 0.2824g + 0.413$.

b) Give an interpretation of the value of the gradient. [1]

c) Justify the use of linear regression in this case. [1]

Total Marks .... / 11

Calculating Probabilities

1 A group of 100 people were asked about the pets they own:

- 58 people owned a cat.
- 42 people owned a dog.
- 20 people owned a rabbit.
- 35 people owned both a cat and a dog.

- 4 people owned both a cat and a rabbit.
- 6 people owned both a rabbit and a dog.
- 2 people owned a cat, a dog and a rabbit.

a) Complete the Venn diagram. [4]

b) Find the probability that a person selected at random did not own any pets. [1]

c) Find the probability that a person selected at random owned a cat only. [1]

d) Find the probability that a person selected at random owned a rabbit, given that they owned a dog. [2]

e) Lorna states that owning a cat and a dog are mutually exclusive events. Is Lorna correct? Explain your answer. [1]

f) Rosemary states that owning a cat and owning a rabbit are independent events. Is Rosemary correct? Explain your answer. [2]

2 Dave Chigwell plays darts. The probability that he hits the bullseye on his first attempt is 0.7.
If he hits the bullseye on his first attempt, the probability that he hits it on his second attempt is 0.8.
If he misses the bullseye on his first attempt, the probability that he hits it on his second attempt is 0.4.

a) On a separate sheet of paper, draw a tree diagram to represent this situation. [3]

b) Find the probability that Dave hits the bullseye once in his two attempts. [3]

Total Marks _____ / 17

Statistical Distributions

1 In a hospital, 40% of patients listen to the hospital radio station. Twenty patients are chosen at random.

    a) State the distribution of the random variable. [2]

    b) Find the probability that at least one-quarter of these patients listen to the hospital radio. [2]

2 A fair, four-sided spinner is numbered 1 to 4.

    a) Construct a table giving the probability distribution of X. [2]

    b) State the name of this type of distribution. [1]

3 A manufacturer produces a large number of toy mice for cats. It is known from previous records that 30% of the toy mice are blue. A random sample of 25 toys is taken.

    a) Find the probability that the sample contains no more than 10 blue toy mice. [3]

    b) Find the probability that the sample contains at least 15 blue toy mice. [3]

4 Twelve counters are numbered 1 to 12 and placed in a bag. One counter is chosen at random and the number, N, recorded.

    a) Write down one condition which needs to be met for the variable N to be modelled as a discrete uniform distribution. [1]

    b) Find the probability that:

        i) $N = 6$ [1]

        ii) N is prime [1]

        iii) $3 \leqslant N < 8$ [1]

5 A random variable $X \sim B(20, 0.8)$. Find the probability that:

a) $X = 4$ [2]

b) $X \geqslant 14$ [2]

Total Marks .... / 21

Hypothesis Testing

1 A four-sided spinner is numbered 1 to 4. Renee believes the spinner is biased towards the number 3. The spinner is spun 40 times and the random variable X represents the number of 3s obtained.

a) State the distribution of the random variable X. [1]

Renee records fifteen 3s.

b) Test, at the 5% significance level, if this is enough evidence to support Renee's claim. State your hypotheses clearly. [7]

2 Find the critical region for the test statistic X, given that X follows the binomial distribution $X \sim B(25, p)$. In each case, test at the 5% significance level.

a) $H_0: p = 0.2$, $H_1: p > 0.2$ [3]

b) $H_0: p = 0.3$, $H_1: p \neq 0.3$ [3]

c) $H_0: p = 0.8$, $H_1: p > 0.8$ [3]

3 John plays the tin-can game at the fairground. The probability of him knocking down all the tin cans is 0.3. John decides to go home and practise to see if he can improve his chances of winning. He returns to the fairground and plays 20 games. He knocks all of the tin cans down 11 times.

Test, at a 10% significance level, if there is enough evidence to support the theory that practice improved his chances of winning. [7]

Total Marks .... / 24

Quantities and Units in Mechanics

1 A resultant force of 12 N acts on a 15 kg mass. Find its increase in velocity after five seconds. [3]

2 Particle A lies on the x-axis at the point $x = -5$. Particle B lies at $x = 10$. If A has a velocity of 0.45 ms⁻¹, and B has a velocity of 0.3 ms⁻¹, calculate the position on the x-axis where they will meet. [6]

3 An object is thrown down from the Eiffel Tower and after 3.2 seconds its velocity is 34 ms⁻¹. Find the speed with which the object is thrown. [3]

> **Total Marks** .... / 12

Displacement–time and Velocity–time Graphs

1 A cyclist accelerates at a constant rate from rest, reaching a speed of 4 ms⁻¹ after three seconds. He maintains this speed for a further four seconds, before decelerating to rest at a rate of 0.8 ms⁻².

   a) Draw a velocity–time graph, illustrating the journey of the cyclist. [3]

   b) Find the time at which the cyclist completed half of his journey. [4]

2 Albert stands 2 m away from a wall. He rolls a ball towards the wall at 4 ms⁻¹. The ball rebounds with speed 3.5 ms⁻¹. Albert moves back 50 cm and stops the ball when it reaches him.

   a) On a separate sheet of paper, draw a displacement–time graph for the ball. [3]

b) Find the average speed of the ball. [5]

Total Marks _____ / 15

Kinematics with Constant and Variable Acceleration

1 A particle travels in a straight line with acceleration $0.75\,\text{ms}^{-2}$. After travelling $26\,\text{m}$, its velocity is $8\,\text{ms}^{-1}$. Find the time taken for its journey. [4]

2 A particle moves in a straight line, such that after time t its displacement s is given by the formula

$$s = \frac{t^3}{3} - \frac{3t^2}{2} + \frac{5t}{2} + 5$$

a) Show that the particle's velocity is always positive. [4]

b) Find the time(s) during which the particle is decelerating. [3]

3 A particle, travelling along a horizontal, is initially at a point O. Its acceleration is given as $a = 2t$ at time t and, after three seconds, its velocity is $-3\,\text{ms}^{-1}$. Find the time when the particle returns to its starting point. [7]

Total Marks _____ / 18

Forces and Newton's Laws 1

1 A particle's velocity changes from $(-2\mathbf{i} - 10\mathbf{j})\,\text{ms}^{-1}$ to $(-\mathbf{i} + 8\mathbf{j})\,\text{ms}^{-1}$ over 10 seconds due to a constant force **F**. Find the distance it travels from its original position. [5]

2 A child of mass 35 kg is inside a lift. Given that the lift is accelerating vertically downwards at a rate of $2 \, \text{ms}^{-2}$, find the force exerted by the child on the floor of the lift. [4]

3 A small car of mass 150 kg is driven along a straight road. The car experiences a constant resistive force. Given that the car moves with constant speed when the driver exerts a forward force of 30 N, calculate the acceleration of the car when the driving force is increased to 50 N. [4]

> Total Marks _____ / 13

Forces and Newton's Laws 2

1 A particle is projected vertically upwards with speed $u \, \text{ms}^{-1}$. T seconds later it is travelling vertically downwards with speed $\frac{u}{2} \, \text{ms}^{-1}$.

a) Find T in terms of u and g. [3]

b) Find the particle's displacement, s, at time $t = T$. [3]

2 A ball is projected upwards with speed $15 \, \text{ms}^{-1}$ from the top of a cliff, 100 m high. Find how long it will take for the ball to reach the bottom of the cliff. [5]

3 Two particles, of mass M kg and 5 kg ($M > 5$), are joined by light string, hanging over a smooth pulley. Initially they are at rest at the same horizontal level. After being released, they both accelerate at a rate of $0.75 \, \text{ms}^{-2}$.

a) Find the value of M. [6]

b) Assuming that the particles are released at a maximum height of 2 m from the floor, and that the 5 kg mass never hits the pulley, find the height above the floor reached by the 5 kg mass. [4]

> Total Marks _____ / 21

Mixed Questions

Pure Mathematics

1 In this question a and b are numbers where $a = b + 2$. The sum of a and b is equal to the product of a and b. Show that a and b are not integers. [3]

2 Calculate the following integrals:

a) $\displaystyle\int_0^4 (3x+2)\,\mathrm{d}x$ [2]

b) $\displaystyle\int_{-1}^1 (24-6t^2)\,\mathrm{d}t$ [2]

c) $\displaystyle\int_{4.7}^{5.1} (3.2s+14.6)\,\mathrm{d}s$ [2]

d) $\displaystyle\int_1^2 \left(\frac{2}{x^4}+\frac{3}{5x^2}\right)\mathrm{d}x$ [2]

3 A rectangular piece of card has side lengths $(2x + 3)$ cm and $(3x - 4)$ cm.

a) Write and simplify an expression for the perimeter of the card. [1]

b) Write and simplify an expression for the area of the card. [1]

c) A child cuts out a rectangle of area $3\,\text{cm}^2$ from the card. Write a factorised expression representing the remaining area of the card. [2]

4 A(−1, −3), B(8, 4) and C(10, 0) form the vertices of a triangle.

a) Find the length of line segment BC. [1]

b) Point D lies on the line that passes through A and is perpendicular to line segment BC. Find the equation of line AD. [3]

c) Find the length of line AD. [3]

d) Hence, or otherwise, find the area of triangle ABC. [1]

5 A closed cylindrical can has a volume of 350 cm³. Find the dimensions of the can which would minimise the surface area. [7]

6 $f(x) = 3x^2 + 8kx - 4k$ has two real roots.

Show that k is within the set $\left\{ k : k < -\dfrac{3}{4} \right\} \cup \{ k : k > 0 \}$. [4]

7 $f(x) = -x(x - 3)(x - 2)(x + 1)$

a) Sketch the graph of $y = f(x)$. [4]

b) The point A(−2, 0) lies on the graph of $y = f(x + a)$. Find the possible values of a. [4]

8 Show that $\dfrac{30x^4 y^{\frac{3}{2}}}{15x^2 y^{\frac{7}{2}}}$ can be written in the form $ax^m y^n$. [1]

9 Solve the equation $\ln(5x) = 0$.

Hence, or otherwise, solve the inequality $\ln(5x) < 0$. [4]

10 Given $\mathbf{a} = 4\mathbf{i} + 5\mathbf{j}$ and $\mathbf{b} = \mathbf{i} + 3\mathbf{j}$, find a vector of magnitude 10 acting in the direction of $\mathbf{a} - 2\mathbf{b}$. [6]

11 The net of a tin of baked beans is shown. The length of the metal sheet from which the tin is cut is 26 cm. The surface area of the tin is $112\pi \, \text{cm}^2$.

Find the volume of the tin. [6]

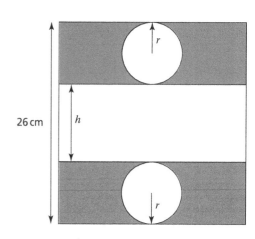

26 cm

12 a) The equation of curve A is of the form $y = \dfrac{a}{x}$

Write down the equation of curve A given that the point (1, 9) lies on the curve. [1]

b) Given the equation of curve B is $y = 2x^2 + 5x - 6$, show that the points of intersection of curves A and B are the solutions to the equation $(x + 3)(x + 1)(2x - 3) = 0$. [3]

c) Hence, find the points of intersection of curves A and B. [2]

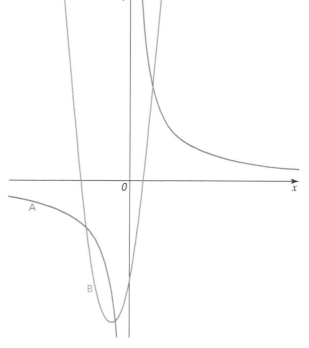

13 Solve the equation $4(\log_2 x)^2 - 20\log_2 x + 21 = 0$, giving your answers in the form $p\sqrt{2}$, where $p \in \mathbb{R}$. [8]

14 Sketch the graph of $y = -2\sin\theta$ in the interval $-2\pi \leqslant \theta \leqslant 2\pi$. Clearly label all points of intersection with the x and y-axes. [3]

15 If $s = \sin\theta$, $c = \cos\theta$ and $t = \tan\theta$

   a) Express $\sqrt{(1 - s^2)}$ in terms of c. [2]

   b) Express $\dfrac{1 - c^2}{c^2}$ in terms of t. [2]

16 Evaluate:

   a) $\displaystyle\int \frac{1}{p^2}\, dp$ [1]

   b) $\displaystyle\int \left(2 - \frac{1}{t^2}\right) dt$ [1]

   c) $\displaystyle\int \frac{2 - 2\sqrt{x}}{x^2}\, dx$ [1]

   d) $\displaystyle\int (x - 4)(x + 3)\, dx$ [1]

17 A gardener is enclosing a vegetable patch along the side of her house. She plans to use 20 m of fencing to enclose an area of 40 m².

Find the possible dimensions of the enclosure to the nearest centimetre. [5]

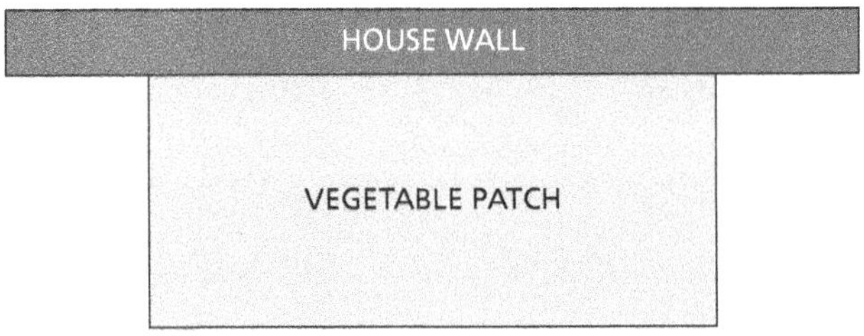

HOUSE WALL

VEGETABLE PATCH

18 Given that f(−2) = 0, fully factorise f(x) = $6x^3 + 5x^2 - 17x - 6$. [5]

19 A(−6, 14), B(10, 2) and C(c, 8) lie on a circle.

  a) Find the value of c given that $c < 0$. [5]

  b) Find the equation of the circle. [4]

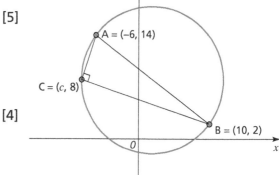

20 Given the vectors $\mathbf{a} = \begin{pmatrix} \lambda \\ 3 \end{pmatrix}$ and $\mathbf{b} = \begin{pmatrix} 1 \\ -3 \end{pmatrix}$, find the value(s) of λ such that $|2\mathbf{a} + 3\mathbf{b}| = \sqrt{34}$. [8]

21 **a)** Sketch the graph of f(x) = (x − 3)(x − 1)(x + 2). [3]

  b) Sketch the graph of $y = f\left(\dfrac{1}{4}x\right)$. [3]

22 **a)** Show that $\dfrac{h^3 + 3}{2h} = \dfrac{1}{2}h^2 + \dfrac{3}{2h}$ [2]

b) Hence find $\dfrac{dy}{dx}$ [3]

c) Hence find $\dfrac{d^2y}{dx^2}$ [2]

23 Show that the sum of any three consecutive multiples of 3 is also a multiple of 3. [3]

24 A student carries out a radioactive decay experiment. He knows that the amount of radioactive substance N is related to time t by the equation $N = N_0 e^{-\lambda t}$, where N_0 and λ are constants. He further finds that when $t = 5$, $N = 200$ and when $t = 7$, $N = 180$.

Given this information, find the value of N_0 and the value of λ to 3 significant figures. [5]

25 Solve $\sin 5\theta = 0$, $180° \leqslant \theta \leqslant 540°$. [3]

26 Given that $\dfrac{dy}{dx} = 7x^3 + 5x^2 - 7x + 1$, find y. [3]

Total Marks .... / 138

Statistics and Mechanics

1 During a norovirus outbreak, 5% of the population of a city was affected on a given day. The manager of a call centre employs 50 people and 10 of them were absent, claiming to have the virus.

a) Using a 5% significance level, find the critical region that would enable the manager to test whether or not there is evidence that the proportion of people who have the virus at her call centre is higher than the proportion of people in the city with the virus. [5]

b) State the conclusion the manager would come to and give a reason for your answer. [2]

2 A particle accelerates from rest at $3.5\,ms^{-2}$ for 2.5 seconds, then decelerates at a constant rate for a further 10.5 seconds until stationary.

a) Sketch a velocity–time diagram to illustrate the particle's journey. [4]

b) Find the total distance travelled by the particle. [2]

3 Each day, a quality control manager measures the number of defective items which are rejected during the production process. The results for the month of July are shown.

Number of faults	0	1	2	3	4	5	6
Number of days	12	10	8	0	0	0	1

a) Calculate the mean number of faults per day. [2]

b) Calculate the standard deviation for this data. [2]

c) Comment whether or not the manager should use these values to make conclusions about the reliability of the production line. Explain your answer. [2]

4 An object of mass 5 kg changes velocity from 5 ms⁻¹ to −5 ms⁻¹ in five seconds.
Calculate the magnitude of the force on the object. [5]

5 Forces F_1, F_2 and F_3 act on a particle that remains in equilibrium. Given $F_1 = (2p\mathbf{i} + q\mathbf{j})$ N, $F_2 = (-q\mathbf{i} + p\mathbf{j})$ N and $F_3 = ((5q - 1)\mathbf{i} + 2p\mathbf{j})$ N, find the value of p and the value of q. [5]

6 The scatter diagram shows temperature plotted against sales in a London shopping centre.

Daniel concludes that the temperature impacts on how much money people spend in the shops. Comment on his conclusion. [2]

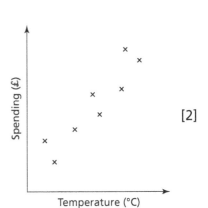

7 The discrete random variable X has the probability distribution below:

x	−2	−1	0	1	2
P($X = x$)	0.1	0.3	0.3	k	$2k$

a) Find the value of k. [2]

b) Find P($-1 < x \leq 2$). [1]

8 A random variable $Y \sim B(8, 0.1)$.

Find the probability:

a) $Y \leq 2$ [1]

b) $Y \geq 7$ [1]

c) $Y < 4$ [1]

9 A particle moves along a straight line with constant acceleration. After five seconds it has travelled a distance of 42.5 m. After another five seconds, it has travelled a further 117.5 m.

a) Find its initial velocity. [4]

b) Find its acceleration. [1]

10 A 6 kg mass and a 3 kg mass are joined by a piece of light string. The 6 kg mass rests on a horizontal table, with the 3 kg mass hanging vertically from one end of the table, passing over a smooth pulley. The system is initially at rest, then released. During the subsequent motion, the 6 kg mass encounters resistances totalling 15 N.

a) i) Find the acceleration of the system. [5]

　　ii) Find the tension in the string. [1]

After two seconds, the string snaps.

b) Find the further distance the 6 kg mass will travel before stopping. [5]

Mixed Questions

11 **a)** Give one advantage and one disadvantage of using a census. [2]

b) Give one advantage and one disadvantage of using a sample. [2]

12 The table shows information about the height of flowers in a field, measured to the nearest centimetre. Calculate estimates for the lower quartile, median and upper quartile.

Height (h cm)	Frequency
$80 < h \leqslant 90$	5
$90 < h \leqslant 95$	10
$95 < h \leqslant 100$	26
$100 < h \leqslant 105$	8
$105 < h \leqslant 110$	1

[4]

13 The histogram shows information about the time taken by a group of students to complete a mathematical puzzle. 104 students took part.

Find the probability that a student completed the puzzle in a time between 14 and 20 seconds.

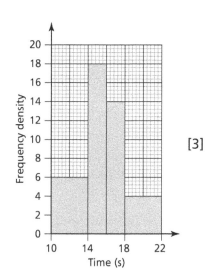

[3]

14 The number of patients experiencing complications after an operation in hospitals across England is 15%. A hospital takes a sample of 30 and finds that 10 patients experienced complications.

Test, at a 1% significance level, if this is enough evidence to suggest that the proportion of patients experiencing complications at this hospital is higher than the national average. [7]

Total Marks .... / 71

Answers

Pages 6–23 Revise Questions

Page 7 Quick Test

1. $\dfrac{x^5}{2}$ 2. $10 + 8\sqrt{2}$ 3. $\dfrac{9 + 3\sqrt{5}}{4}$

Page 9 Quick Test

1. $3x^3 - 5xy + 18x - 4y$ 2. $2x^3 + x^2 - 7x - 6$
3. $(3x + 2)(x - 4)$ 4. $(2x - 5)(3x - 4)$
5. $2x(x + 3)(x + 1)$

Page 11 Quick Test

1. $1 + 24x + 264x^2 + 1760x^3$ 2. $\dfrac{x - 1}{2(x + 2)}$

3. a) $f(-1) = 2(-1)^3 + (-1)^2 - 7(-1) - 6 = 0$
   b) $2x^2 - x - 6$ c) $(x + 1)(2x + 3)(x - 2)$

Page 13 Quick Test

1. $x = 4 - \sqrt{14}$ and $x = 4 + \sqrt{14}$

2. $x = \dfrac{-1 - \sqrt{19}}{3}$ and $x = \dfrac{-1 + \sqrt{19}}{3}$

3.
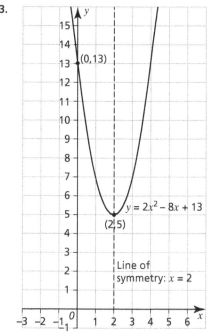

Page 15 Quick Test

1. $x = 2, y = -1$ 2. $x = -3, y = 1$
3. $x = -6, y = 16$ and $x = 2, y = 0$

Page 17 Quick Test

1. $x \leqslant 7$ 2. $\left\{ x : -\dfrac{1}{4} \leqslant x < 0 \right\}$ 3. $\{x : x < 1\} \cup \{x : x > 4\}$

Page 19 Quick Test

1.
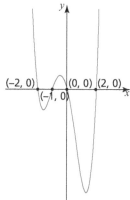

2.
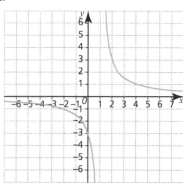

3. a: $k = \dfrac{1}{2}$ d: $k = \dfrac{3}{4}$

Page 21 Quick Test

1. a) $y = x(x - 1)(x + 2)(x + 4) \ (= x^4 + 5x^3 + 2x^2 - 8x)$
   b) $y = (x - 1)(x - 2)(x + 1)(x + 3) + 1 = x^4 + x^3 - 7x^2 - x + 7$

2. a)
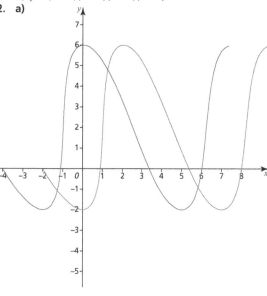

   b)
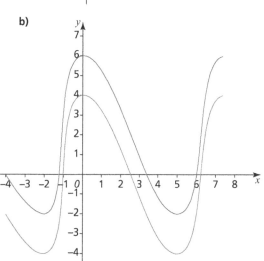

Page 23 Quick Test

1. and 2.
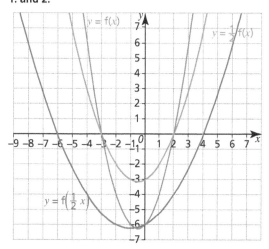

3. and 4.

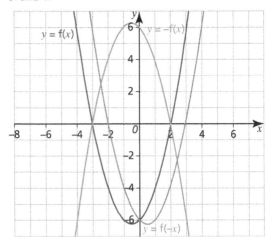

Page 24: Indices and Surds

1. a) $6x^6y^8$ **[1]**   b) $3^4 \times m^{4 \times 4} \times n^{\left(\frac{3}{2} \times 4\right)} = 81m^{16}n^6$ **[1]**

    c) $\frac{(16k^6j^2)^{\frac{1}{2}}}{4j} = \frac{\sqrt{16} \times k^{6 \times \frac{1}{2}} \times j^{2 \times \frac{1}{2}}}{4j}$ **[1]** $= \frac{4k^3j}{4j} = k^3$ **[1]**

2. a) $\left(3 + \sqrt{2}\right) - \left(1 - \sqrt{8}\right) = \left(3 + \sqrt{2}\right) - \left(1 - \sqrt{4} \times \sqrt{2}\right)$

    $= 3 + \sqrt{2} - 1 + 2\sqrt{2} = 2 + 3\sqrt{2}$ **[1]**

    b) $\left(2 + \sqrt{7}\right)\left(3 + \sqrt{7}\right) = 6 + 2\sqrt{7} + 3\sqrt{7} + 7 = 13 + 5\sqrt{7}$ **[1]**

3. a) $\frac{2}{4 + \sqrt{7}} = \frac{2\left(4 - \sqrt{7}\right)}{\left(4 + \sqrt{7}\right)\left(4 - \sqrt{7}\right)}$ **[1]** $= \frac{8 - 2\sqrt{7}}{16 - 7} = \frac{8 - 2\sqrt{7}}{9}$ **[1]**

    b) $\frac{3 + \sqrt{10}}{3 - \sqrt{10}} = \frac{\left(3 + \sqrt{10}\right)\left(3 + \sqrt{10}\right)}{\left(3 - \sqrt{10}\right)\left(3 + \sqrt{10}\right)}$ **[1]**

    $= \frac{9 + 6\sqrt{10} + 10}{9 - 10} = \frac{19 + 6\sqrt{10}}{-1} = -19 - 6\sqrt{10}$ **[1]**

Page 24: Manipulating Algebraic Expressions

1. $a = 7, b = -22, c = 3$ $a \times c = 21$
    $7k^2 - 22k + 3 = 7k^2 - 21k - k + 3 = 7k(k - 3) - 1(k - 3)$
    $= (7k - 1)(k - 3)$ **[1]**

2. $(x + 2)(3x - 8) - 2x(y + 4) - 3y = 3x^2 - 8x + 6x - 16 - 8x - 2xy - 3y$ **[1]**
    $= 3x^2 - 2xy - 10x - 3y - 16$ **[1]**

3. $\frac{(2m + 4) + (4m - 8)}{2} \times (2m - 1)$ **[1]** $= \frac{6m - 4}{2} \times (2m - 1)$

    $= (3m - 2)(2m - 1)$ **[1]**
    $= 6m^2 - 7m + 2$ **[1]**

Page 25: Expanding and Dividing Polynomials

1. Using Pascal's triangle or the binomial expansion, the coefficient of x^4 is $5ab^4$. Substituting in $a = 3, 5 \times 3 \times b^4 = 240$ **[1]**
    $b^4 = \frac{240}{15}$ so $b = \sqrt[4]{16}$ **[1]** $\Rightarrow b = 2$ or $b = -2$ **[1]**

    Don't forget that roots of even powers have two possible answers: one positive and one negative.

2. $f(-3) = 0$ (because $(x + 3)$ is a factor of $f(x)$)
    Substituting $x = -3$ into $f(x)$:
    $f(-3) = a \times (-3)^3 + (-3)^2 - 13(-3) + 6$ **[1 for substitution]**
    Remember that $f(-3) = 0$; replace $f(-3)$ with 0.
    $0 = -27a + 9 + 39 + 6$ **[1 for setting equal to 0]**
    $-27a = -54$, so $a = 2$ **[1]**

Page 25: Quadratic Equations

1. Complete the square of $ax^2 + bx + c = 0$
    $ax^2 + bx + c = 0$
    $\quad ax^2 + bx = -c$
    $\quad a(x^2 + \frac{b}{a}x) = -c$ **[1]**

    $a\left(x^2 + \frac{b}{a}x + \left(\frac{b}{2a}\right)^2\right) = -c + \left(a \times \left(\frac{b}{2a}\right)^2\right)$

$a\left(x + \frac{b}{2a}\right)^2 = -c + \frac{b^2}{4a}$ **[1]**

$a\left(x + \frac{b}{2a}\right)^2 + \left(c - \frac{b^2}{4a}\right) = 0$ **[1]**

The value of 'p' is 'a', the value of 'q' is $\frac{b}{2a}$ and the value of 'r' is $\left(c - \frac{b^2}{4a}\right)$

2. $2x^2 + 2x - 4 = 0$ Solve f(x) = 0 for x to find the roots.

    $(x - 1)(x + 2) = 0$, so $x = 1, x = -2$ **[1]**
    $y = 2x^2 + 2x - 4$

    Complete the square of $y = f(x)$ to find the turning point.

    $y = 2\left(x + \frac{1}{2}\right)^2 - \frac{9}{2}$ **[1 for $\frac{1}{2}$, 1 for $\frac{9}{2}$]**

    y-intercept is –4 **[1]**

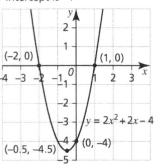

[1 for fully correct graph]

3. The y-intercept is the initial height. Substitute the values of the minimum point into the function $f(x) = p(x - q)^2 + r$ and expand to find the value of the y-intercept. Remember the value of 'p' is the value of 'a' when $f(x) = ax^2 + bx + c$.

    $f(x) = 2(x - 0.6)^2 + 1$ **[1 for p = 2, 1 for q = 0.6 and r = 1]**
    $f(x) = 2(x^2 - 1.2x + 0.36) + 1$ **[1 for expansion]**
    $f(x) = 2x^2 - 2.4x + 1.72$
    The initial height is 1.72 m **[1]**

Page 26: Simultaneous Equations

1. $4x + 2y = 2$ Rearrange $4x + 2y = 2$ to express y in terms of x.
    $2y = 2 - 4x$, so $y = 1 - 2x$ **[1]**

    Substitute $y = 1 - 2x$ into $y = 2x^2 - x$ and solve for x.

    $2x^2 - x = 1 - 2x$
    $2x^2 + x - 1 = 0$, so $(2x - 1)(x + 1) = 0$ **[1]**
    $2x - 1 = 0, x = \frac{1}{2}$ or $x + 1 = 0, x = -1$ **[1]**

    Substitute $x = \frac{1}{2}$ and $x = -1$ into either equation to find the corresponding y-values.

    $y = 1 - \left(2 \times \frac{1}{2}\right) = 0$ and $y = 1 - (2 \times -1) = 3$

    The solutions are $x = -1, y = 3$ and $x = \frac{1}{2}, y = 0$ **[1]**

2. Substitute $y = x - 1$ into $2x^2 - xy - 3y^2 = 1$
    $2x^2 - x(x - 1) - 3(x - 1)^2 = 1$
    $2x^2 - x^2 + x - 3(x^2 - 2x + 1) = 1$ **[1]**
    $2x^2 - x^2 + x - 3x^2 + 6x - 3 = 1$
    $-2x^2 + 7x - 4 = 0$ **[1]**
    Using the quadratic formula,

    $x = \frac{-b \pm \sqrt{b^2 - 4ac}}{2a} = \frac{-7 \pm \sqrt{7^2 - (4 \times (-2) \times (-4))}}{(2 \times -2)} = \frac{7 \pm \sqrt{17}}{4}$ **[1]**

    Substitute the x-values into the linear equation, $y = x - 1$.

    $y = \frac{7 + \sqrt{17}}{4} - 1 = \frac{3 + \sqrt{17}}{4}$

    $y = \frac{7 - \sqrt{17}}{4} - 1 = \frac{3 - \sqrt{17}}{4}$ **[1]**

The solutions are $x = \dfrac{7 + \sqrt{17}}{4}$, $y = \dfrac{3 + \sqrt{17}}{4}$ and $x = \dfrac{7 - \sqrt{17}}{4}$,

$y = \dfrac{3 - \sqrt{17}}{4}$ [1 – must have solutions paired correctly]

Page 26: Inequalities

1.
Rearrange to $6x^2 + 7x - 3 < 0$ and find the x-intercepts of
$y = 6x^2 + 7x - 3$

$6x^2 + 7x - 3 = 0$, so $(2x + 3)(3x - 1) = 0$
$2x + 3 = 0$, $x = -\dfrac{3}{2}$ or $3x - 1 = 0$, $x = \dfrac{1}{3}$ **[1]**

Sketch the graph, showing
the x-intercepts **[1]** and
identify the values of x for

which $y < 0$, $x > -\dfrac{3}{2}$ and

$x < \dfrac{1}{3}$ **[1]**

$\left\{ x: -\dfrac{3}{2} < x < \dfrac{1}{3} \right\}$ **[1]**

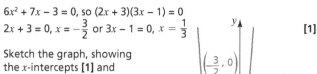

2. $-3(x - 1) + 4 > 5$
$-3x + 7 > 5$, so $-3x > -2$ **[1]**
$x < \dfrac{2}{3}$ **[1]**

Reverse the inequality symbol when dividing by a negative number.

3. $-\dfrac{3}{x} \geqslant 2$

$-3x \geqslant 2x^2$ Multiply both sides by x^2.   **[1]**

$-2x^2 - 3x \geqslant 0$ Rearrange so that one side is 0 and sketch the graph.

$-2x^2 - 3x = 0$ Find the x-intercepts.

$-x(2x + 3) = 0$, $x = 0$ and $x = -\dfrac{3}{2}$ **[1 for each value]**

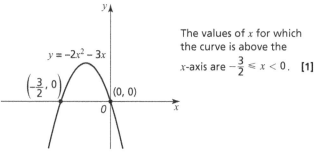

The values of x for which
the curve is above the
x-axis are $-\dfrac{3}{2} \leqslant x < 0$. **[1]**

Page 27: Sketching Curves

1. $(x^4 + x^3 - 2x^2) \div (x - 1) = x^3 + 2x^2$ **[1]**

$$x - 1 \overline{\smash{\big)}\,\begin{array}{l} x^3 + 2x^2 \\ x^4 + x^3 - 2x^2 \\ \underline{-(x^4 - x^3)} \downarrow \\ 2x^3 - 2x^2 \\ \underline{-(2x^3 - 2x^2)} \\ 0 \end{array}}$$

Factorise $x^3 + 2x^2 = x^2(x + 2)$ **[1]**
Then, $f(x) = x^4 + x^3 - 2x^2 = (x^2)(x + 2)(x - 1)$ intersects the
x-axis at $(0, 0)$ (repeated), $(-2, 0)$ and $(1, 0)$ (when $f(x) = 0$) **[1]**
$f(x)$ intersects the y-axis at $(0, 0)$ (when $x = 0$, $f(0) = 0$)

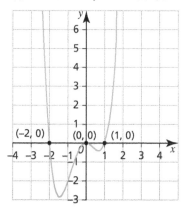

[1]

2. Factorise $2x^3 - x^2 - x = x(2x^2 - x - 1) = x(2x + 1)(x - 1)$ **[1]**
The cubic equation intersects the x-axis at $(-0.5, 0)$ $(0, 0)$ and
$(1, 0)$. **[1]**
The y-intercept is $(0, 0)$. **[1]**

The curve $y = \dfrac{2}{x^2}$ has asymptotes at $x = 0$ and $y = 0$.

Calculate a few points and plot the curve.

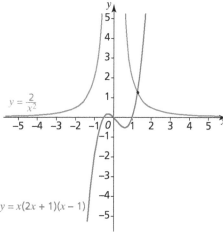

[2 marks: 1 for each curve]

3. The x-intercepts give the value of x when $y = 0$, so the equation
is $y = (x - 2)(x - 3)(x - 5)$ **[1]**
The y-intercept is the value of y when $x = 0$,
$y = (0 - 2)(0 - 3)(0 - 5) = -30$ **[1]**

Page 28: Translating Graphs

1. $f(x) = 2x^2 - 3x - 2 = (2x + 1)(x - 2)$ **[1]**
The roots of $f(x)$ are when $f(x) = 0$
$(2x + 1)(x - 2) = 0$
$x = -\dfrac{1}{2}$, $x = 2$ **[1]**
The function $f(x + 2)$ is a translation of $f(x)$ by -2 units in the
x-direction, so the roots are $x = -\dfrac{5}{2}$ and $x = 0$. **[1]**

You could also use the equation of $f(x + 2)$ and find the roots.

2. a) $y = f(x) - 2$

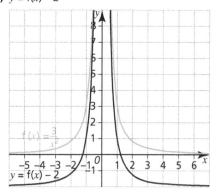

[1]
Asymptotes at $x = 0$ and $y = -2$ **[1]**

b) $y = f(x - 2)$

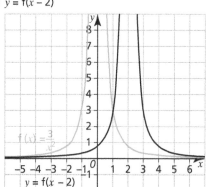

[1]
Asymptotes at $x = 2$ and $y = 0$ **[1]**

Page 29: Stretching and Reflecting Graphs

1. **a)** A transformation of the form $y = f(ax)$ maps each point (x, y) on to $\left(\frac{1}{a}x, y\right)$ so $(3, -20)$ becomes $(15, -20)$ [1]

   b) A transformation of the form $y = af(x)$ maps each point (x, y) on to (x, ay) so $(3, -20)$ becomes $(3, -100)$ [1]

2.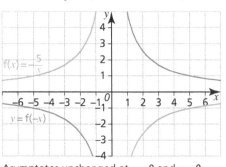

   [1 for general shape; 1 for correct quadrants]

   Asymptotes unchanged at $y = 0$ and $x = 0$ [1]

3.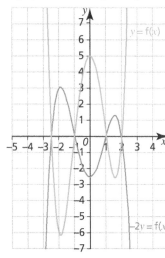

   [1 for correct x-intercepts; 1 for correct y-intercept; 1 for reflection; 1 for stretch by a factor of $\frac{1}{2}$]

Page 31 Quick Test

1. $y = -\frac{2}{3}x + 1$

2. Midpoint: $\left(\frac{9}{2}, 10\right)$ or $(4.5, 10)$ Distance: 5 units

3. $-2x + 5y + 15 = 0$
4. $-5x + 4y +$ any value $= 0$ or $y = \frac{5}{4}x +$ any value

Page 33 Quick Test

1. Centre $(5, 1)$, $r = 2\sqrt{3}$ 2. $y = -\frac{4}{3}x - \frac{32}{3}$

Page 35 Quick Test

1.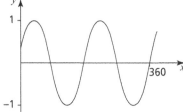

2. $\cos 40°$ 3. $120.2°$ 4. $23.8°$

Page 37 Quick Test

1. $\cos\theta$ 2. $\frac{1}{2}\cos^2\theta$

3. $\frac{5}{13}$ 4. $\frac{\sin\theta}{\cos\theta} + \frac{\cos\theta}{\sin\theta} = \frac{\cos^2\theta + \sin^2\theta}{\sin\theta\cos\theta} = \frac{1}{\sin\theta\cos\theta}$

Page 39 Quick Test

1. $390°, 510°$ 2. $\frac{2}{3}\pi, \frac{4}{3}\pi$ 3. $51°$

Page 40: Equations of Straight Lines

1. **a)** $m_{L1} = \frac{-7 - (-1)}{2 - (-2)} = -\frac{6}{4} = -\frac{3}{2}$ [1]

   $m_{L2} = \frac{5 - (-1)}{7 - (-2)} = \frac{6}{9} = \frac{2}{3}$ [1]

   $-\frac{3}{2} \times \frac{2}{3} = -1$ [1]

   b) Find the equations of L_1 and L_2.

   L_1: $y - (-1) = -\frac{3}{2}(x - (-2))$, so $y = -\frac{3}{2}x - 4$ [1]

   L_2: $y - (-1) = \frac{2}{3}(x - (-2))$, so $y = \frac{2}{3}x + \frac{1}{3}$ [1]

2. $m_{AB} = \frac{2 - (-3)}{-2 - (-5)} = \frac{5}{3}$ $m_{BC} = \frac{7 - 2}{1 - (-2)} = \frac{5}{3}$ [1]

   Since $m_{AB} = m_{BC}$ and B lies on both AB and BC, then A, B and C are collinear. [1]

3. $m_{AB} = \frac{4 - (-2)}{5 - (-3)} = \frac{3}{4}$ $m_{DC} = \frac{-6 - 0}{0 - 8} = \frac{3}{4}$ [1 for both]

   $m_{BC} = \frac{0 - 4}{8 - 5} = -\frac{4}{3}$ $m_{AD} = \frac{-6 - (-2)}{0 - (-3)} = -\frac{4}{3}$ [1 for both]

   AB is parallel to DC and BC is parallel to AD.
   $\frac{3}{4} \times -\frac{4}{3} = -1$, so AB is perpendicular to BC and to AD. Therefore ABCD is a rectangle.
   [1 for stating parallel, perpendicular and rectangular]

   $d_{AB} = \sqrt{(5 - (-3))^2 + (4 - (-2))^2} = 10$ [1]

   $d_{BC} = \sqrt{(8 - 5)^2 + (0 - 4)^2} = 5$ [1]

   Area $= 10 \times 5 = 50$ units^2 [1]

Page 41: Circles

1.   Complete the square on $x^2 + y^2 + 2fx + 2gy + c = 0$

   $x^2 + 2fx + y^2 + 2gy = -c$
   $(x + f)^2 + (y + g)^2 = -c + f^2 + g^2$ [1]
   Comparing to $(x - a)^2 + (y - b)^2 = r^2$, the centre is $(-f, -g)$ [1]
   and the radius is $\sqrt{f^2 + g^2 - c}$ [1]

2. **a)** The perpendicular from centre to chord bisects the chord.

   $m_{AB} = \frac{y_2 - y_1}{x_2 - x_1} = \frac{7 - (-3)}{1 - (-9)} = 1$, so the gradient of its perpendicular bisector is -1. [1]

   The midpoint of AB is
   $\left(\frac{x_1 + x_2}{2}, \frac{y_1 + y_2}{2}\right) = \left(\frac{-9 + 1}{2}, \frac{-3 + 7}{2}\right) = (-4, 2)$ [1]

   The diameter is a line that passes through $(-4, 2)$ and has a gradient of -1.
   $(y - y_1) = m(x - x_1)$
   $(y - 2) = -1(x - (-4))$, so $y = -x - 2$ [1]

   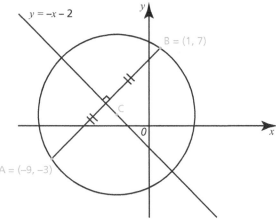

   b) $y = -x - 2$
   $y = -(-3) - 2$, so $y = 1$ [1]
   The centre is $(-3, 1)$.

The size of the radius is the distance between the centre and either given point on the circumference. Using point B:

$$d = \sqrt{(x_2 - x_1)^2 + (y_2 - y_1)^2}$$

$$d = \sqrt{(-3 - 1)^2 + (1 - 7)^2} = \sqrt{52}$$ **[1]**

Thus, the equation of the circle is $(x + 3)^2 + (y - 1)^2 = 52$ **[1]**

Page 42: Sine, Cosine & Tangent Functions, and Sine & Cosine Rule

1.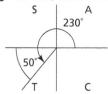

cos is negative in quadrant 3.

Identify the angle 50 **[1]** $\cos 250° = -\cos 50°$ **[1]**

2. a) 280° **[1]**   b) +80 **[1]**   c) 0.985 (3 s.f.) **[1]**
   d) (10, 1)   **[2]**

   The graph would normally cross the x-axis at 0, 180 and 360, therefore a shift of 80° to the left.

3. a) $\sin(32.6) = \dfrac{BD}{8.7}$, $BD = 4.7$ **[1]**

   $(BC)^2 = 4.7^2 + 9.8^2 - 2 \times 4.7 \times 9.8 \times \cos(100) = 134.13$ **[1]**
   $BC = 11.6$ **[1]**

   b) $\dfrac{1}{2} \times 4.7 \times 9.8 \times \sin(100)$ **[1]**

   $\dfrac{1}{2} \times 8.7 \times 4.7 \times \sin(57.4)$ **[1]**

   Summation of two triangle areas **[1]**   39.8 (3 s.f.) **[1]**

Page 43: Trigonometric Identities

1. $\cos^2 A = \dfrac{7}{11}$, $\sin A = \dfrac{\sqrt{4}}{\sqrt{11}}$ **[1]**

   $\tan A = \dfrac{\sin A}{\cos A} = \dfrac{\frac{2}{\sqrt{11}}}{\frac{\sqrt{7}}{\sqrt{11}}} = \dfrac{2}{\sqrt{11}} \times \dfrac{\sqrt{11}}{\sqrt{7}} = \dfrac{2}{\sqrt{7}}$ **[1]** $= \dfrac{2\sqrt{7}}{7}$ **[1]**

2. $(\cos^2 x - \sin^2 x)(\cos^2 x + \sin^2 x)$ The difference of two squares. **[1]**
   Using $\cos^2 x + \sin^2 x = 1$ **[1]**
   $\cos^2 x - \sin^2 x$ **[1]**

3. $1 + 2\sin\theta + \sin^2\theta + \cos^2\theta$ **[1]**
   $1 + 2\sin\theta + 1$ **[1]**
   $2 + 2\sin\theta$ **[1]**; $2(1 + \sin\theta)$ **[1]**

Page 43: Solving Trigonometric Equations

1. $2\sin^2\theta - \sin\theta - 1 = 0$ **[1]**
   $(2\sin\theta + 1)(\sin\theta - 1) = 0$ Replace $\cos^2\theta$ with $1 - \sin^2\theta$.
   $\sin\theta = \dfrac{-1}{2}$ and $\sin\theta = 1$ **[1]**

   $\theta = -30°$ or $90°$ To find the second **[1]**
   $\theta = 210°, 330°$ and $90°$ solution, $180 - $ first; -30 **[1]**
   is outside the range.

2. $\dfrac{\sin 2\theta}{\cos 2\theta} = 1$, $\tan 2\theta = 1$ **[1]**

   $\tan^{-1}(1) = 45°$ (not in range) **[1]**
   $45 - 180 = -135$ **[1]**, $-135 - 180 = -315$ **[1]**

3. $2\tan^2\theta - 3\tan\theta = 0$ **[1]**

   Factorise rather than divide by $\tan\theta$, otherwise you will lose solutions.

   $\tan\theta(2\tan\theta - 3) = 0$ **[1]**
   $\tan^{-1}(0) = 0$ $\theta = 0, 180, 360$ **[1]**
   $\tan^{-1}\left(\dfrac{3}{2}\right) = 56.3$, $\theta = 56.3, 236.3$ **[1]**

4. $0 \leqslant \theta \leqslant 2\pi$, $\dfrac{\pi}{3} \leqslant \left(\theta + \dfrac{\pi}{3}\right) \leqslant \dfrac{7\pi}{3}$

   $\sin\left(\theta + \dfrac{\pi}{3}\right) = \pm\dfrac{1}{\sqrt{2}}$ **[1]**

   $\sin^{-1}\left(\dfrac{1}{\sqrt{2}}\right) = \dfrac{\pi}{4}$ (out of range) **[1]** $\pi - \dfrac{\pi}{4} = \dfrac{3\pi}{4}, 2\pi + \dfrac{\pi}{4} = \dfrac{9\pi}{4}$ **[1]**

   $\sin^{-1}\left(\dfrac{-1}{\sqrt{2}}\right) = -\dfrac{\pi}{4}$ (out of range) **[1]** $\pi - \dfrac{-\pi}{4} = \dfrac{5\pi}{4}, -\dfrac{\pi}{4} + 2\pi = \dfrac{7\pi}{4}$ **[1]**

5. a) $\dfrac{1 - \sin^2\theta}{\sin\theta(1 + \sin\theta)}$ **[1]** $\dfrac{(1 - \sin\theta)(1 + \sin\theta)}{\sin\theta(1 + \sin\theta)}$ **[1]**   The difference of two squares.

   $= \dfrac{1 - \sin\theta}{\sin\theta} = $ RHS **[1]**

b) $\dfrac{1 - \sin\theta}{\sin\theta} = 2$, $1 - 3\sin\theta = 0$ **[1]**

$\sin^{-1}\dfrac{1}{3} = 19.5$ **[1]**
$180 - 19.5 = 160.5$ **[1]**

Pages 44–49 Review Questions

Page 44: Indices and Surds

1. a) $\left(2a^4 b^{\frac{3}{2}}\right)^{-2} = \dfrac{1}{\left(2a^4 b^{\frac{3}{2}}\right)^2}$ **[1]** $= \dfrac{1}{2^2 \times a^{4 \times 2} \times b^{\frac{3}{2} \times 2}} = \dfrac{1}{4a^8 b^3}$ **[1]**

   b) $3a^2 b^3 \times 4a^3 \div ab^5 = \dfrac{12a^5 b^3}{ab^5} = \dfrac{12a^4}{b^2}$ **[1]**

   c) $3^{\frac{1}{2}} + 3^{\frac{5}{2}} - 3^{\frac{3}{2}} = 3^{\frac{1}{2}} + 3^{2\frac{1}{2}} - 3^{1\frac{1}{2}} = 3^{\frac{1}{2}} + 3^2 \times 3^{\frac{1}{2}} - 3 \times 3^{\frac{1}{2}}$ **[1]**

   $= \sqrt{3} + 9\sqrt{3} - 3\sqrt{3} = 7\sqrt{3}$ **[1]**

   Fractional indices can be split up in this way since $a^{mn} = a^m \times a^n$

2. a) $\dfrac{1}{12}$ **[1]**

   b) $\sqrt[3]{5^6 \times 15^{-9}} = \left(5^6 \times 15^{-9}\right)^{\frac{1}{3}} = 5^{6 \times \frac{1}{3}} \times 15^{-9 \times \frac{1}{3}} = 5^2 \times 15^{-3}$ **[1]**

   $= 5^2 \times 5^{-3} \times 3^{-3} = 5^{2-3} \times 3^{-3} = 5^{-1} \times 3^{-3} = \dfrac{1}{5 \times 3^3} = \dfrac{1}{135}$ **[1]**

3. a) $\dfrac{2}{3 - \sqrt{5}} = \dfrac{2(3 + \sqrt{5})}{(3 - \sqrt{5})(3 + \sqrt{5})}$ **[1]**

   $= \dfrac{6 + 2\sqrt{5}}{9 - 5} = \dfrac{6 + 2\sqrt{5}}{4}$

   $= \dfrac{3 + \sqrt{5}}{2}$ **[1]**

   b) $\dfrac{\sqrt{7}}{4 + \sqrt{7}} = \dfrac{\sqrt{7}(4 - \sqrt{7})}{(4 + \sqrt{7})(4 - \sqrt{7})}$ **[1]**

   $= \dfrac{4\sqrt{7} - 7}{16 - 7} = \dfrac{4\sqrt{7} - 7}{9}$ **[1]**

Page 44: Manipulating Algebraic Expressions

1. a) $(3a - 3)(2a - 2) - (2a + 1)(a - 1) = 6a^2 - 6a - 6a + 6 -$
   $(2a^2 - 2a + a - 1)$ **[1]**
   $= 6a^2 - 12a + 6 - (2a^2 - a - 1)$
   $= 6a^2 - 2a^2 - 12a + a + 6 + 1$
   $= 4a^2 - 11a + 7$ **[1]**

   b) $4a^2 - 11a + 7 = 4a^2 - 4a - 7a + 7$
   $= 4a(a - 1) - 7(a - 1) = (4a - 7)(a - 1)$ **[1]**

2. $3x - 12x^3 = 3x(1 - 4x^2)$ **[1]** $= 3x(1 + 2x)(1 - 2x)$ **[1]**

3. $2x^2 - 4x - 6 = 2(x^2 - 2x - 3)$ **[1]** $= 2(x + 1)(x - 3)$ **[1]**

Page 45: Expanding and Dividing Polynomials

1.
$$\begin{array}{r} x^3 - 2x^2 - 5x + 6 \\ x + 4 \overline{\smash{)}\, x^4 + 2x^3 - 13x^2 - 14x + 24} \\ \underline{- (x^4 + 4x^3)} \\ -2x^3 - 13x^2 \\ \underline{- (-2x^3 - 8x^2)} \\ -5x^2 - 14x \\ \underline{- (-5x^2 - 20x)} \\ 6x + 24 \\ \underline{- (6x + 24)} \\ 0 \end{array}$$
[1 for attempting division]
[1 for quotient correct]

$x^4 + 2x^3 - 13x^2 - 14x + 24 = (x + 4)(x^3 - 2x^2 - 5x + 6)$ **[1]**
$a = 1, b = -2, c = -5, d = 6$ **[1]**

Remember to answer the question by writing the given polynomial in the required format and don't stop after the division.

2. a) The first four terms are

$\dbinom{10}{0} 2^{10}(-x)^0 + \dbinom{10}{1} 2^9(-x)^1 + \dbinom{10}{2} 2^8(-x)^2 + \dbinom{10}{3} 2^7(-x)^3$

[2 for all four terms correct; 1 for at least two correct]
$= 1024 - 5120x + 11520x^2 - 15360x^3$ **[1]**

b) To find an approximation of 1.97^{10}, substitute $x = 0.03$

$1024 - (5120 \times 0.03) + (11520 \times 0.03^2) - (15360 \times 0.03^3)$

[1 for substitution of 0.03]

$= 880.3533$ **[1]**

Page 45: Quadratic Equations

1.

Complete the square of $ax^2 + bx + c = 0$ and rearrange to solve for x.

$ax^2 + bx + c = 0$, so $ax^2 + bx = -c$

$a\left(x^2 + \dfrac{b}{a}x\right) = -c$ **[1]**

$a\left(x^2 + \dfrac{b}{a}x + \left(\dfrac{b}{2a}\right)^2\right) = -c + \left(a \times \left(\dfrac{b}{2a}\right)^2\right)$

$a\left(x + \dfrac{b}{2a}\right)^2 = -c + \dfrac{b^2}{4a}$ **[1]**

$a\left(x + \dfrac{b}{2a}\right)^2 = \dfrac{-4ac}{4a} + \dfrac{b^2}{4a}$ **[1]**

$\left(x + \dfrac{b}{2a}\right)^2 = \dfrac{b^2 - 4ac}{4a^2}$ **[1]**

$x + \dfrac{b}{2a} = \dfrac{\sqrt{b^2 - 4ac}}{2a}$ **[1]**

$x = \dfrac{\sqrt{b^2 - 4ac}}{2a} - \dfrac{b}{2a}$, so $x = \dfrac{-b \pm \sqrt{b^2 - 4ac}}{2a}$ **[1]**

2. Let $m = x^4$, then $y = x^8 - 18x^4 + 32 = m^2 - 18m + 32$ **[1]**
The roots are the values of x when $y = 0$.
Factorise $0 = m^2 - 18m + 32 = (m - 2)(m - 16)$
Substitute $m = x^4$, then $0 = (x^4 - 2)(x^4 - 16)$ **[1]**
$x^4 - 2 = 0$, $x = \pm\sqrt[4]{2}$

or $x^4 - 16 = 0$, $x = \pm\sqrt[4]{16} = \pm2$ **[1 for solving for x in either bracket]**

The roots are $x = -2$, $x = 2$, $x = -\sqrt[4]{2}$ and $x = \sqrt[4]{2}$
[1 (must have all four roots)]
3. $b^2 - 4ac = 4^2 - (4 \times (-2) \times (-5))$ **[1]**
$= 16 - 40 = -24$, so $b^2 - 4ac < 1$ **[1]**, meaning the function has no real roots so does not intersect the x-axis.

Page 46: Simultaneous Equations
1. Using elimination:

$2x + 3y = 8$ (1)
$3x - 2y = 4$ (2)
$6x + 9y = 24$ (3)   Multiply (1) by 3.

$6x - 4y = 8$ (4)   Multiply (2) by 2.   **[1 for multiplying both equations]**

$13y = 16$ Subtract (3) − (4)

$y = \dfrac{16}{13}$ **[1]**

$2x + 3y = 8$, $2x + \left(3 \times \dfrac{16}{13}\right) = 8$, $x = \dfrac{28}{13}$ **[1 for substitution]**

The solution is $x = \dfrac{28}{13}, y = \dfrac{16}{13}$ **[1]**

2. Substitute $y = 2x + 1$ into $(x + 1)^2 + (y - 1)^2 = 4$
$(x + 1)^2 + (2x + 1 - 1)^2 = 4$
$x^2 + 2x + 1 + 4x^2 = 4$ **[1]**
$5x^2 + 2x - 3 = 0$
$(x + 1)(5x - 3) = 0$ **[1]**
$x + 1 = 0$, $x = -1$ $5x - 3 = 0$, $x = \dfrac{3}{5}$

Substitute the x-values to find the corresponding y-values.

$x = -1$, $y = (2 \times -1) + 1 = -1$

$x = \dfrac{3}{5}$, $y = \left(2 \times \dfrac{3}{5}\right) + 1 = \dfrac{11}{5}$

The solutions are $x = -1$, $y = -1$ **[1]** and $x = \dfrac{3}{5}$, $y = \dfrac{11}{5}$ **[1]**

3. a)

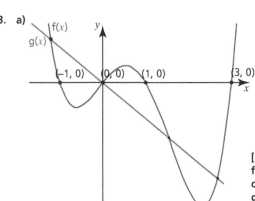

[1 for g(x); 1 for x-intercepts of f(x); 1 for general shape of f(x)]

b) Expanding the brackets on f(x):
$x(x + 1)(x - 1)(x - 3) = (x^2 + x)(x - 1)(x - 3) = (x^3 - x)(x - 3)$
$= x^4 - 3x^3 - x^2 + 3x$ **[1]**
Setting f(x) equal to g(x):
$x^4 - 3x^3 - x^2 + 3x = -2x$, so $x^4 - 3x^3 - x^2 + 5x = 0$ **[1]**
f(x) intersects g(x) at 4 points, so there are 4 real solutions. **[1]**

Page 46: Inequalities
1.

Find the points of intersection of f(x) and g(x).

$3x^2 - 5x - 2 = 2x + 4$, so $3x^2 - 7x - 6 = 0$ **[1]**

$(3x + 2)(x - 3) = 0$, $x = -\dfrac{2}{3}$ or $x = 3$ **[1]**

Looking at the graph, f(x) > g(x) when $x < -\dfrac{2}{3}$ or when $x > 3$,

$\left\{x : x < -\dfrac{2}{3}\right\} \cup \{x : x > 3\}$ **[1]**

2.

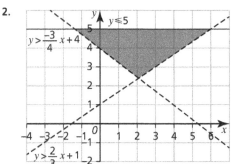

[1 for all three lines correct; 1 for dotted and solid line correct; 1 for correct shading]

Page 47: Sketching Curves
1. $-x(x + 1)(x - 1)(x - 3) = (x - 2)(x - 4)$ **[1]**
$-x(x + 1) = x^2 - x$
$(-x^2 - x)(x - 1) = x^3 + x^2 - x^2 + x$
          $= -x^3 + x$ **[1]**
$(-x^3 + x)(x - 3) = x^4 + 3x^3 + x^2 - 3x$ **[1]**
$(x - 2)(x - 4) = x^2 - 6x + 8$ **[1]**
$-x^4 + 3x^3 + x^2 - 3x = x^2 - 6x + 8$ **[1]** (or equivalent equation, must be expanded)
The solutions are $x = 1.34$ and $x = 3.04$ **[1]**
[These are estimates; any values of x where $1 \leqslant x \leqslant 1.5$ and $2.8 \leqslant x \leqslant 3.2$ are acceptable]

2. The table does not show direct proportion, as the ratio of $\dfrac{y}{x}$ is not constant. **[1]**

Time (min), x	0	3	6	9	12	15
Temp. (°C), y	0	7	25	65	98	100
Ratio $\dfrac{y}{x}$	0	2.33	4.17	7.22	8.17	6.67

[1 for showing that the ratio is not the same]

Direct proportion is defined as $y = kx$ for some constant k. If two variables are in direct proportion, the ratio of $\dfrac{y}{x}$ will always be k.

Page 48: Translating Graphs
1. The x-coordinate has increased by 5 and the y-coordinate has decreased by 2. **[1]** Hence, the transformation is $y = f(x - 5) - 2$
[2 marks: 1 for each value a and b]

2. The graph of $f(x) = x^2$ has a turning point at $(0, 0)$; the graph of $y = f(x - 2) - 3$ represents a translation of $+2$ units in the x-direction and -3 units in the y-direction, so the turning point is $(2, -3)$. **[1]**

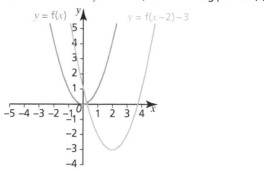

[1]

3. $y = f(x + a)$ shows a translation of $y = f(x)$ by -1 in the x-direction, so $a = 1$ **[1]**
Hence the equation is $f(x + 1) = (x - 2)(x + 1)(x + 2)$ **[1]**

Page 49: Stretching and Reflecting Graphs

1. a) A transformation of the form $y = f(ax)$ maps each point (x, y) on to $\left(\frac{1}{a}x, y\right)$, so $(-2, 5)$ becomes $(-4, 5)$. **[1]**

b) $4y = f(x)$ can be rearranged to $y = \frac{1}{4}f(x)$ **[1]**

A transformation of the form $y = af(x)$ maps each point (x, y) on to (x, ay) so $(-2, 5)$ becomes $\left(-2, \frac{5}{4}\right)$ **[1]**

2. a) and b)

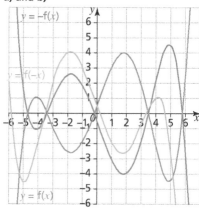

[1 for general shape; 1 for correct transformation]

Page 51 Quick Test

1.

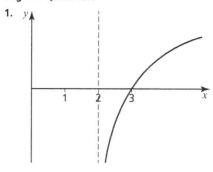

2. $x = \dfrac{1 + \ln\left(\frac{2}{3}\right)}{4}$

3. $\dfrac{e^{-1} - 2}{3}$

4. $4e^2 \approx 29.6$

Page 53 Quick Test
1. 7 **2.** $x = -0.644$ **3.** $x = \sqrt{10}$ or $x = 1000$

Page 55 Quick Test
1. Plot x on x-axis, $\ln y$ on y-axis.
2. Plot $\log_2 x$ on x-axis, y on y-axis.
3. Plot x^2 on x-axis, $\frac{y}{x}$ on y-axis.
4. Plot $\dfrac{\log_2 x}{x}$ on x-axis, $\dfrac{y}{x}$ on y-axis.

Page 57 Quick Test
1. $30x^5 + 8$ **2.** $-3x^{-2} - 16x^{-3}$
3. $84x^3 - 90x^5$ **4.** $x = \frac{1}{9}, x = 1$

Page 59 Quick Test
1. $x < \frac{9}{2}$ **2.** $x = \frac{1}{2}$, maximum **3.** $y = 7x - 36, 7y = -x + 48$

Page 60: Exponentials and Logarithms
1.

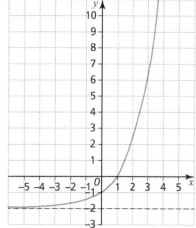

[1 for shape; 1 for asymptote at $y = -2$; 1 for crossing y-axis at $y = -1$; 1 for crossing x-axis at $x = 1$]

2. a) $\dfrac{dy}{dx} = ke^{kx}$ **[1]** $= ke^{k\left(\frac{1}{k}\right)}$ **[1]** $= ke$ **[1]**

b) $y - e = ke\left(x - \dfrac{1}{k}\right) \Rightarrow y = kex$ **[2]**

3. $\dfrac{x}{2} = \ln\left(\dfrac{1}{2}\right)$ **[1]**; $x = 2\ln\left(\dfrac{1}{2}\right)$ **[1]**; $x = -2\ln 2$ **[1]**

Page 60: Laws of Logarithms
1. $\log_{10} 3^{4x} = \log_{10} 45$ **[1]**
$4x\log_{10} 3 = \log_{10} 45$ **[1]**
$x = \dfrac{\log_{10} 45}{4\log_{10} 3} = 0.866$ **[1]**

2. $\log_3 4 - \log_3 36 = \log_3\left(\dfrac{4}{36}\right)$ **[1]**

$= \log_3\left(\dfrac{1}{9}\right)$ **[1]** $= \log_3(3^{-2})$ **[1]**

$= -2\log_3 3$ **[1]** $= -2$ **[1]**

3. a) $(2^x + 2)(2^x - 3) = 0$ **[1]**
$2^x = -2 \Rightarrow$ no solutions **[1]**
so $2^x = 3 \Rightarrow x = \dfrac{\log_{10} 3}{\log_{10} 2}$ **[2]**

b) $\log_{10} a + \dfrac{1}{\log_{10} a} = \dfrac{10}{3}$ **[1]**

$(\log_{10} a)^2 + 1 = \dfrac{10}{3}\log_{10} a$

$3(\log_{10} a)^2 - 10\log_{10} a + 3 = 0$ **[1]**
$(3\log_{10} a - 1)(\log_{10} a - 3) = 0$ **[1]**
$\log_{10} a = \frac{1}{3}$ or $\log_{10} a = 3$ **[1]**

$a = 10^{\frac{1}{3}}$ **[1]** or $a = 1000$ **[1]**

Page 61: Modelling Using Exponentials and Logarithms
1. Initial value is when $t = 0$, i.e. $P = A$ **[1]**
So when $P = 2A$, $2A = Ae^{0.2t}$
$2 = e^{0.2t}$ **[1]**; $\ln 2 = 0.2t$ **[1]**
$t = 5\ln 2 = 3.47$ months **[1]**
2. Substituting values into the equation gives:
$45 = A \times 10^{2k}$ and $45\,000 = A \times 10^{8k}$ **[1]**
Dividing equation 2 by equation 1:
$10^{6k} = \dfrac{45000}{45} = 1000$ **[1]**
$6k = 3$ so $k = \dfrac{1}{2}$ **[1]**
Substituting into equation 1 gives: $45 = A \times 10$
$\Rightarrow A = 4.5$ **[1]**
3. $y = ax^n \Rightarrow \log_{10} y = \log_{10} a + n\log_{10} x$ **[2]**
Gradient $= n = 1.48$ **[1]** $\Rightarrow n = 1.5$ **[1]**
y-intercept $= \log_{10} a = -0.6$ **[1]** $\Rightarrow a = 10^{-0.6} = 0.25$ **[1]**

Page 62: Differential of a Polynomial
1. a) $f(x) = x + 9x^{-1}$ **[1]**
$f'(x) = 1 - 9x^{-2}$ **[1]**
b) $1 - 2x^{-2} = 0$, $x^2 = 2$ **[1]**; $x = \pm 3$ **[1]**

2. a) $y = 2x^3 + x^{\frac{1}{2}} + \dfrac{x^2}{x^2} + \dfrac{2x}{x^2}$ **[1]**

$y = 2x^3 + x^{\frac{1}{2}} + 1 + 2x^{-1}$ Write in power form and simplify. **[1]**

$\dfrac{dy}{dx} = 6x^2 + \dfrac{1}{2}x^{\frac{-1}{2}} - 2x^{-2}$ **[1]**

b) $6(5)^2 + \dfrac{1}{2}(5)^{\frac{-1}{2}} - 2(5)^{-2}$ **[1]** = 150 (3 s.f.) **[1]**

3. $y = 6x^2 + 12x - 10x - 20$ **[1]**; $6x^2 + 2x - 20$ **[1]**

$\dfrac{dy}{dx} = 12x + 2$ **[2]**

$\dfrac{d^2y}{dx^2} = 12$ **[1]**

4. $f'(x) = 12x^2 - 24x$ **[1]**; $f''(x) = 24x - 24$ **[1]**

Page 62: Stationary Points, Tangents and Normals
1. a) $3x^2 - 12x + 9 = 0$ **[1]**; $(3x - 3)(x - 3) = 0$ **[1]**
$x = 1, y = 4$ **[1]**; $x = 3, y = 0$ **[1]**

b) $\dfrac{d^2y}{dx^2} = 6x - 12$ **[1]**

$x = 1, \dfrac{d^2y}{dx^2} = -6$ (negative $\therefore$ maximum)

$x = 3, \dfrac{d^2y}{dx^2} = 6$ (positive $\therefore$ minimum) **[1 for both statements]**

2. $\dfrac{dy}{dx} = 2x + 4$ **[1]**
$x = 3$, gradient = 10 **[1]**
$y - 18 = 10(x - 3) \Rightarrow y = 10x - 12$ **[1]**

3. $y = 2 - 2x^{-2}$
$\dfrac{dy}{dx} = 4x^{-3}$ **[1]**
$m = \dfrac{-1}{16}$ Remember the normal is perpendicular. **[1]**
$y - \dfrac{15}{8} = 16(x - -4)$ **[1]**
$8y - 128x - 497 = 0$ **[1]**

4. a) $\dfrac{dy}{dx} = 4x^3 - 24x^2 - 124x + 144$ **[2 (1 for reduction of a power)]**

b) When $x = 1$, $\dfrac{dy}{dx} = 4(1)^3 - 24(1)^2 - 124(1) + 144 = 0 \therefore$ shown **[1]**
$(x - 1)(4x^2 - 20x - 144)$ **[1]**

Remember to divide the polynomial by $(x - 1)$.

$(x - 9)(x + 4) = 0, x = 9, x = -4$ **[1]**
c)

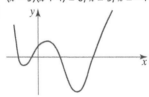

[1 for shape; 1 for positive y-intercept]

5. $f'(x) = 10x + 12$ **[1]**
$10x + 12 > 0$ **[1]**, so $x > \dfrac{-6}{5}$ **[1]**

Pages 64–67 Review Questions

Page 64: Equations of Straight Lines
1. Finding the gradient: $m = \dfrac{y_2 - y_1}{x_2 - x_1} = \dfrac{12 - 9}{11 - 5} = \dfrac{1}{2}$ **[1]**

Substitute into $y = mx + c$ and rearrange.

$9 = \left(\dfrac{1}{2} \times 5\right) + c$, so $c = \dfrac{13}{2}$ **[1]**

$y = \dfrac{1}{2}x + \dfrac{13}{2}$ **[1]**
$2y = x + 13 \Rightarrow -x + 2y - 13 = 0$ **[1]**

2. Since the lines are perpendicular, $m_{L2} = -\dfrac{3}{2}$ **[1]**

The lines meet at point A, so by using simultaneous equations
$\dfrac{2}{3}x - 4 - 7 = -\dfrac{3}{2}(x - 10)$ **[1 for simultaneous equations]**

$\dfrac{2}{3}x - 11 = -\dfrac{3}{2}x + 15$, so $\dfrac{2}{3}x + \dfrac{3}{2}x = 15 + 11$

$\dfrac{13}{6}x = 26$ **[1 for attempting to rearrange and solve]**

$x = 12$ **[1]**

$y = \left(\dfrac{2}{3} \times 12\right) - 4$, so $y = 4$ **[1]**

The coordinates of A are (12, 4).

Page 64: Circles
1. $x^2 + y^2 + 6x - 2y + 10 = 10$
$(-6)^2 + 0^2 + (6 \times -6) - (2 \times 0) + 10 = 10$, as required. **[1]**

Rearrange and complete the square.

$x^2 + 6x + y^2 - 2y = 0$
$(x + 3)^2 + (y - 1)^2 = 10$ **[1]**
Centre C(-3, 1)
$m_{AC} = \dfrac{1 - 0}{-3 - (-6)} = \dfrac{1}{3}$, so $m_{tangent} = -3$ **[1]**
$y - 0 = -3(x - (-6))$
$y = -3x - 18$ **[1]**

2. $r = \sqrt{(x_2 - x_1)^2 + (y_2 - y_1)^2} = \sqrt{(-5 - (-2))^2 + (12 - 8)^2} = 5$ **[1]**
$(x + 2)^2 + (y - 8)^2 = 25$ **[1]**

3. Midpoint of AB: $\left(\dfrac{-2 + 6}{2}, \dfrac{5 + 9}{2}\right) = (2, 7)$
[1 for midpoint of AB and midpoint of AC]

Gradient of AB: $m = \dfrac{9 - 5}{6 - (-2)} = \dfrac{1}{2}$
[1 for gradient of AB and gradient of AC]

Gradient of perpendicular bisector is -2.
$y - y_1 = m(x - x_1)$
$y - 7 = -2(x - 2) \Rightarrow y = -2x + 11$ **[1]**
Midpoint of AC: $\left(\dfrac{-2 + 3}{2}, \dfrac{5 + 0}{2}\right) = \left(\dfrac{1}{2}, \dfrac{5}{2}\right)$

Gradient of AC: $m = \dfrac{0 - 5}{3 - (-2)} = -1$
Gradient of perpendicular bisector is 1.
$y - y_1 = m(x - x_1)$; $y - \dfrac{5}{2} = 1\left(x - \dfrac{1}{2}\right) \Rightarrow y = x + 2$ **[1]**

Circumcentre: $x + 2 = -2x + 11$
[1 for setting up simultaneous equations]
$x = 3, y = 5$ **[1 for x and y correct]**
Radius: $r = \sqrt{(x_2 - x_1)^2 + (y_2 - y_1)^2} = \sqrt{(-2 - 3)^2 + (5 - 5)^2} = 5$ **[1]**
Equation of the circumcircle is $(x - 3)^2 + (y - 5)^2 = 25$ **[1]**

Page 65: Sine, Cosine & Tangent Functions, and Sine & Cosine Rule
1. a)

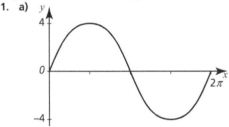

[1 for correct shape; 1 for 4, -4 and 2π correctly shown]

b) $(\pi, 4)$ **[1]**; $\left(\dfrac{3\pi}{2}, -4\right)$ **[1]**

2. $\dfrac{5\pi}{8} - \dfrac{\pi}{8} = \dfrac{\pi}{2}$ or $\dfrac{9\pi}{8} - \dfrac{5\pi}{8} = \dfrac{\pi}{2}$ **[1]** $\therefore a = 2$ **[1]**

Shift $\dfrac{\pi}{4}$ left, therefore $b = \dfrac{\pi}{4}$ **[1]**

3. $0.5 \times 30 \times 40 \times \sin(x) = 400$ **[1]**; $\sin(x) = \dfrac{2}{3}$ **[1]** $x = 41.8°$ **[1]**
4. $2.4^2 = 1.2^2 + 3^2 - 2 \times 1.2 \times 3 \times \cos(x)$; $\cos(x) = \dfrac{13}{20}$ **[1]**
$x = 49.5$ (3 s.f.) **[1]**
$A = 0.5 \times 1.2 \times 3 \times \sin(49.5)$ **[1]** = 1.37 cm² (3 s.f.) **[1]**

Page 66: Trigonometric Identities
1. $\cos^2x - (1 - \cos^2x)$ **[1]**; $2\cos^2x - 1 =$ RHS **[1]**
2. $\sin^2x(\sin^2x + \cos^2x)$ **[1]** $\therefore \sin^2x$ **[1]**
3. $\dfrac{1}{\cos x} - \dfrac{\cos^2 x}{\cos x}$ **[1]**; $\dfrac{\sin^2 x}{\cos x}$ **[1]**; $\sin x \times \dfrac{\sin x}{\cos x} = \sin x \tan x =$ RHS **[1]**

4. a) $\sin x = \sqrt{1 - \cos^2 x} = \sqrt{1 - \frac{9}{16}}$ [1] $= \pm\frac{\sqrt{7}}{4}$

As x is reflex and $\cos x$ is positive, quadrant 4 $\therefore \sin x = -\frac{\sqrt{7}}{4}$ [1]

b) $\tan x = \frac{\sin x}{\cos x} = \frac{\sqrt{7}}{4} \times \frac{16}{9}$ [1] $= \frac{4\sqrt{7}}{9}$

As x is reflex and $\cos x$ is positive, quadrant 4 $\therefore -\frac{4\sqrt{7}}{9}$ [1]

5. $x^2 = \sin^2 x$ or $y^2 = \cos^2 x$ [1]

$x^2 + y^2 = 1$ [1]

Page 67: Solving Trigonometric Equations

1. $\sin x = 1$ or $\cos x = \frac{-3}{5}$ [1]

$x = 90$ [1]

$\cos^{-1}\left(\frac{-3}{5}\right) = 126.9$ [1]

$x = 360 - 126.9 = 233.1$ [1]

2. $\sin(x - 30) = \pm\frac{1}{\sqrt{2}}$ [1]

$\sin^{-1}\left(\frac{1}{\sqrt{2}}\right) = 45$ and $180 - 45 = 135$ [1]

$\sin^{-1}\left(\frac{-1}{\sqrt{2}}\right) = -45$ and $180 - -45 = 225$ (out of range)

$x - 30 = 45$ or 135 so $x = 75$ or 165 [1]

3. a) $(1 - \sin^2 3x) + 4\sin 3x - 1$ [1]

$4\sin 3x - \sin^2 3x$ [1]

$\sin 3x(4 - \sin 3x) = $ RHS [1]

b) $\sin 3x = 0$ or $\sin 3x = 4$ (no solutions) [1]

$0° \leqslant x \leqslant 360°,\ 0° \leqslant 3x \leqslant 1080°$ [1]

$3x = 0, 180, 360, 540, 720, 900, 1080$ [1]

> Look for crossing points and use the symmetry on the graph.

$x = 0, 60, 120, 180, 240, 300, 360$ [1]

Pages 68–77 Revise Questions

Page 69 Quick Test

1. $\frac{7x^3}{3} - \frac{5x^2}{2} + 2x + c$ **2.** $\frac{x^4}{4} - \frac{1}{2}x^{-1} + c$ **3.** $x^3 + x^2 - 2$

Page 71 Quick Test

1. a) $\frac{341}{3}$ **b)** 6   **2.** $\frac{-1}{6}$

Page 73 Quick Test

1. $p = 12\cos 30i + 12\sin 30j = 6\sqrt{3}i + 6j$ **2.** $\frac{\sqrt{5}}{5}i + \frac{2\sqrt{5}}{5}j$

Page 75 Quick Test

1. a) $\overline{AB} = -4i + 12j$, $\overline{BC} = 7i$ **b)** $\sqrt{\frac{160}{49}} = \frac{4\sqrt{10}}{7}$

2. $r = \frac{29}{4}i + \frac{23}{4}j$

Page 77 Quick Test

1. Expand the brackets

2. $(2m + 1)(2n + 1) = 4mn + 2m + 2n + 1 = 2(2mn + m + n) + 1$ $\therefore$ odd

3. $b^2 - 4ac = 9 - 60 = -51$; negative $\therefore$ no real roots

Pages 78–81 Practice Questions

Page 78: Indefinite Integrals

1. a) $\frac{x^5}{5} + \frac{3x^2}{2} + c$ [1]

b) $\frac{4t^{\frac{3}{2}}}{3} + \frac{t^2}{2} + c$ [1]

c) $\frac{x^3}{3} + \frac{3x^2}{2} + 2x + c$ [1]

2. $\frac{6x}{3x^3} - \frac{2}{3x^3}$ [1]; $\frac{2}{x^2} - \frac{2}{3x^3}$ [1]

$2x^{-2} - \frac{2}{3}x^{-3}$ [1]

$= \int 2x^{-2} - \frac{2}{3}x^{-3}dx = -2x^{-1} + \frac{1}{3}x^{-2} + c$ [1]

3. a) $4 - 12\sqrt{x} + 9x$ [1]

$a = 12, b = 9$ [1]

b) $\int\left(4 - 12x^{\frac{1}{2}} + 9x\right)dx$ [1]

$4x - 8x^{\frac{3}{2}} + \frac{9x^2}{2}$ [2]

4. $x^2 + 5x + c$ [1]

$-1 = 3^2 + 5 \times 3 + c$, $c = -25$ [1]

$f(x) = x^2 + 5x - 25$ [1]

5. $s = \frac{3t^2}{2} + 8t^{-1} + c$ [1]

Substitute in the point $t = 1$, $s = \frac{3}{2}$

$s = \frac{3}{2}t^2 + \frac{8}{t} - 8$ [1]

Page 78: Integration to Find Areas

1. a) $\left[\frac{5x^2}{2} + x\right]_1^6$ [1] $= 92.5$ [1]

b) $\left[-\frac{1}{4}q^4 + 50q^2\right]_{20}^{50}$ [1] $= -1417500$ [1]

c) $\left[-x^{-1} + \frac{1}{2}x^{-2}\right]_2^3$ [1] $= \frac{7}{72}$ [1]

2. $\int_0^2 (3x^2 - 2x + 2)\,dx$ [1 for limits of 2 and 0]

$[x^3 - x^2 + 2x]_0^2$ [2] $= 8$ [1]

3. $\int_{-5}^9 t(x)\,dx = \int_{-5}^{-3} t(x)\,dx + \int_{-3}^6 t(x)\,dx + \int_6^9 t(x)\,dx$ [1]

$-4 = \int_{-5}^{-3} t(x)\,dx + 4 - 9$ [1]

$\int_{-5}^{-3} t(x)\,dx = 1$ [1]

4. $\int_1^2 \left(15x^{\frac{1}{2}} + x^{\frac{3}{2}}\right)dx$ [1]

$\left[10x^{\frac{3}{2}} + \frac{2}{5}x^{\frac{5}{2}}\right]_1^2$ [2] $= 20.1$ (3 s.f.) [1]

Page 79: Introduction to Vectors

1. Unit vector is $\frac{7i + 24j}{\sqrt{7^2 + 24^2}}$ [1] $= \frac{7i + 24j}{25}$ [1]

So require $\frac{5(7i + 24j)}{25}$ [1]

$= \frac{7}{5}i + \frac{24}{5}j$ [1]

2. a) $a = 4\cos 60i + 4\sin 60j$ [1]

$= 2i + 2\sqrt{3}j$ [1]

b) $a = 10\cos 150i + 10\sin 150j$ [1]

$= -5\sqrt{3}i + 5j$ [1]

3. Using parallelogram law: [1]

$R^2 = 10^2 + 12^2 - 2 \times 10 \times 12\cos 145°$ [1] $\Rightarrow R = 21.0$ [1]

Using sine rule: [1]

$\frac{\sin\theta}{10} = \frac{\sin 145}{21.0}$ [1] $\Rightarrow \theta = 15.9°$ above the horizontal [1]

Page 80: Vector Geometry

1. Using distance formula:

$5^2 + (k - 3)^2 = 30$ [2]

$(k - 3)^2 = 5$

$k = 3 \pm \sqrt{5}$ [2]

2. Let C be the required point.

$\overline{AB} = \overline{OB} - \overline{OA} = \begin{pmatrix} 45 \\ 80 \end{pmatrix} - \begin{pmatrix} -30 \\ 10 \end{pmatrix} = \begin{pmatrix} 75 \\ 70 \end{pmatrix}$ [2]

$\frac{3}{10}\overline{AB} = \begin{pmatrix} \frac{225}{10} \\ 21 \end{pmatrix}$ [2]

$\overline{OC} = \overline{OA} + \overline{AC} = \begin{pmatrix} -30 \\ 10 \end{pmatrix} + \begin{pmatrix} \frac{225}{10} \\ 21 \end{pmatrix}$ [1] $= \begin{pmatrix} -7.5 \\ 31 \end{pmatrix}$ [1]

3. Let D be the point of intersection of the diagonals.

$\overline{OB} = a + c$ [1]

So $\overline{OD} = \lambda(a + c) = \lambda a + \lambda c$ ...Eqn 1 [1]

$\overline{CA} = -a + c$ [1]

So $\overrightarrow{CD} = \mu(-c + a)$ [1]

$\overrightarrow{OD} = \overrightarrow{OC} + \overrightarrow{CD} = c + \mu(-c + a) = \mu a + (1 - \mu)c$ ...Eqn 2 [1]

Equate coefficients of a: $\lambda = \mu$ [1]

Equate coefficients of b: $1 - \mu = \lambda$ [1]

Solve simultaneously: $\lambda = \mu = \frac{1}{2}$ [1]

So $\overrightarrow{OD} = \frac{1}{2}\overrightarrow{OB}$ and $\overrightarrow{CD} = \frac{1}{2}\overrightarrow{CA}$, so result proven. [2]

Page 81: Proof

1. $(2n + 1)^2 = 4n^2 + 4n + 1$ [1]
   $4(n^2 + n) + 1$ [1] $\therefore$ one more than a multiple of 4 [1]

2. $3n + 1$ [1]
   $(3n + 1)(3m + 1) = 9nm + 3m + 3n + 1$
   $= 3(3nm + m + n) + 1$ and [1]
   $(3nm + m + n)$ is an integer, therefore proved. [1]

3. APC = $180 - 2x$ [1]
   BPC = $180 - (180 - 2x) = 2x$ [1]
   ABC = BPC (isosceles triangle) $\therefore$ ABC = $2x$ [1]

4. $(2a - 1)^2 - (2b + 1)^2 = 4a^2 - 4a + 1 - 4b^2 + 4b - 1$ [1]
   $4a^2 - 4a - 4b^2 + 4b = 4(a^2 - a - b^2 + b)$ [1]
   $\qquad\qquad = 4(a - b)(a + b - 1)$ [1]

5. $(9n^2 + 6n + 1) - (9n^2 - 6n + 1)$ [1]
   $12n = 4(3n)$ [1] $\therefore$ multiple of 4 [1]

6. Gradient of BC $= \frac{5 - 4}{6 - 2} = \frac{1}{4}$; gradient of AD $= \frac{2 - 1}{5 - 1} = \frac{1}{4}$ [1]

   Equal gradients $\therefore$ parallel

   Gradient of AB $= \frac{4 - 1}{2 - 1} = \frac{3}{1}$; gradient of DC $= \frac{5 - 2}{6 - 5} = \frac{3}{1}$ [1]

   Equal gradients $\therefore$ parallel [1]
   Two pairs of parallel sides so parallelogram

   Sketch a diagram.

Page 82: Exponentials and Logarithms

1.
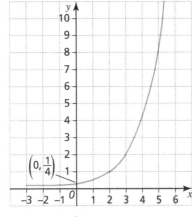

   [1 for shape; 1 for asymptote at $y = 0$; 1 for crossing y-axis at $y = \frac{1}{4}$]

2. $\ln(2x - 3) = \frac{1}{3}$ [1]

   $2x - 3 = e^{\frac{1}{3}}$ [1]; $x = \frac{3 + e^{\frac{1}{3}}}{2}$ [1]

3. $y = e^{6x} - 4e^{3x} + 4$ [2]

   $\frac{dy}{dx} = 6e^{6x} - 12e^{3x}$ [2]

Page 82: Laws of Logarithms

1. $\log_{10}\left(2^{x-1}\right) = \log_{10}\left(\frac{1}{5}\right)$ [1]

   $(x - 1)\log_{10}2 = \log_{10}\left(\frac{1}{5}\right)$ [1]

   $x = \frac{\log_{10}\left(\frac{1}{5}\right)}{\log_{10}2} + 1 = -1.32$ [1]

2. $\log_{10}5 + 2\log_{10}x = \log_{10}10 + \log_{10}8$ [1]
   $\log_{10}5 + 2\log_{10}x = \log_{10}80$ [1]
   $\log_{10}5 + \log_{10}x^2 = \log_{10}80$ [1]
   $\log_{10}(5x^2) = \log_{10}80$ [1]
   $5x^2 = 80$ [1]
   $x^2 = 16 \Rightarrow x = 4$ (only positive solution allowed) [1]

3. $\log_{10}\sqrt{p^4 q^7} = \log_{10}\left(p^4 q^7\right)^{\frac{1}{2}}$ [1] $= \log_{10}\left(p^2 q^{\frac{7}{2}}\right)$ [1]

   $= \log_{10}\left(p^2\right) + \log_{10}\left(q^{\frac{7}{2}}\right)$ [1]

   $= 2\log_{10}p + \frac{7}{2}\log_{10}q$ [1]

Page 83: Modelling Using Exponentials and Logarithms

1. Half-life is when $N = \frac{N_0}{2}$ [1]

   $\frac{N_0}{2} = N_0 e^{-\lambda t}$

   $\frac{1}{2} = e^{-\lambda t}$ [1], so $2 = e^{\lambda t}$ [1]

   $\ln 2 = \lambda t$ [1]

   $t_{\frac{1}{2}} = \frac{\ln 2}{\lambda}$

2. $y = kb^x \Rightarrow \log_{10}y = \log_{10}k + x\log_{10}b$ [2]
   Gradient $= \log_{10}b = 0.40$ [1] $\Rightarrow b = 10^{0.40} = 2.5$ [1]
   y-intercept $= \log_{10}k = 1.18$ [1] $\Rightarrow k = 10^{1.18} = 15$ [1]

3. $y = Axe^{kx}$, so $\frac{y}{x} = Ae^{kx}$ [1]

   $\ln\left(\frac{y}{x}\right) = \ln A + \ln\left(e^{kx}\right)$ [1]; $\ln\left(\frac{y}{x}\right) = \ln A + kx$ [1]

   So plot x on the x-axis [1] and plot $\ln\left(\frac{y}{x}\right)$ on the y-axis. [1]

Page 84: Differential of a Polynomial

1. a) $f'(x) = 3x^2 + 4x + 2$ [2 if fully correct; 1 for one correct power]
   b) $3x^2 + 4x + 2 = 1$
      $3x^2 + 4x + 1 = 0$ [1]
      $(3x + 1)(x + 1) = 0$, $x = -\frac{1}{3}$, $x = -1$ [1]

2. a) $-5x^{-2}; \frac{3}{10}x^{-\frac{4}{5}}$ [2 if fully correct; 1 for reduction of power by 1]

   b) $-5(2)^{-2} + \frac{3}{10}(2)^{-\frac{4}{5}}$ [1]
      -1.08 (3 s.f.) [1]

3. $2x^2 - 4x + 4x - 8$ [1]; $y = 2x^2 - 8$ [1]

   $\frac{dy}{dx} = 4x$ [2 if fully correct; 1 for reduction of power by 1]

   $\frac{d^2y}{dx^2} = 4$ [1]

Page 85: Stationary Points, Tangents and Normals

1. a) $\frac{dy}{dx} = 6x^2 - 16x + 8$ [1]

      $3x^2 - 8x + 4 = 0$, $(3x - 2)(x - 2) = 0$ [1]

      $x = 2$ and $x = \frac{2}{3}$; $(2, 0)$ and $\left(\frac{2}{3}, \frac{64}{27}\right)$ [1]

   b) $\frac{d^2y}{dx^2} = 12x - 16$ [1]

      When $x = 2$, $\frac{d^2y}{dx^2} = 24 - 16$ (positive therefore minimum)

      When $x = \frac{2}{3}$, $\frac{d^2y}{dx^2} = 8 - 16$ (negative therefore maximum)

      [1 for both conclusions]

2. $\frac{dy}{dx} = 6x + 3 - 2x^{-3}$ [1]

   At $x = 1$, $\frac{dy}{dx} = 7$ [1]

   At $x = 2$, $\frac{dy}{dx} = \frac{59}{4}$ [1]

   At $x = 3$, $\frac{dy}{dx} = \frac{565}{27}$ [1]

3. a) $\frac{dy}{dx} = 3x^2 - 10x + 5$ [1]

   b) i) $3x^2 - 10x + 5 = 2$, so $3x^2 - 10x + 3 = 0$
         $(3x - 1)(x - 3) = 0$
         $x = 3$ (P), $x = \frac{1}{3}$ (Q) [1]

      ii) When $x = 3$, $y = -1$ Use the original equation to find y.

         $y + 1 = 2(x + 3)$ [1]
         $y - 2x - 5 = 0$ [1]

      iii) When $x = \frac{1}{3}$, $y = \frac{85}{27}$ Use the original equation to find y.

         $y - \frac{85}{27} = -\frac{1}{2}\left(x - \frac{1}{3}\right)$ [1]

         $54y + 27x - 179 = 0$ [1]

iv) $y - 2x - 5 = 0$

At $x = 0$, $y = 5$ At $y = 0$, $x = -\dfrac{5}{2}$ [1]

> To find crossing points, let $x = 0$ to find the y-intercept and let $y = 0$ to find the x-intercept.

$$\sqrt{\left(\dfrac{5}{2}\right)^2 + 5^2} = \dfrac{5\sqrt{5}}{2}$$ [1]

Pages 86–99 Revise Questions

Page 87 Quick Test

1. A sample is quicker or cheaper or involves less data. A sample may introduce bias and not be a fair reflection of the population.
2. Number all the members of the gym 1 to 400. Use a random number generator, list or table to produce 50 random numbers. Select the member of the population which corresponds to that number.

Page 89 Quick Test

1. Median = 5; Mode = 8; Mean = 5.14 (2 d.p.)
2. Median = 8; Interquartile range = 1
3. Mean = 27.3; Standard deviation = 6.58

Page 91 Quick Test

1.

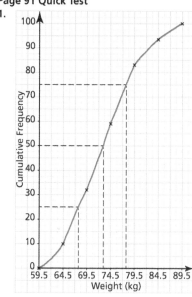

Lower quartile = 67.7–67.9;
Median = 72.9–73.1;
Upper quartile = 77.4–77.6

2.

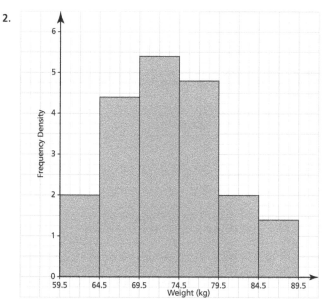

63.8 (i.e. 64 students)

3. The increase in the length of the rod in millimetres per 1 degree increase in temperature.

Page 93 Quick Test

1. 36   2. **a)** 47.5   **b)** 15   **c)** 65

Page 95 Quick Test

1. **a)**

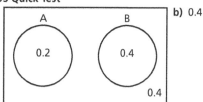

b) 0.4

2. **a)**

6	7	8	9	10	11	12
5	6	7	8	9	10	11
4	5	6	7	8	9	10
3	4	5	6	7	8	9
2	3	4	5	6	7	8
1	2	3	4	5	6	7
	1	2	3	4	5	6

b) $\dfrac{7}{12}$ **3.** $\dfrac{60}{121}$

Page 97 Quick Test

1. 0.336   2. Independent trials; probability of success or failure
3. 0.9452

Page 99 Quick Test

1. $X \geqslant 8$ 2. $X \leqslant 6$ and $X \geqslant 19$

Pages 100–105 Practice Questions

Page 100: Populations and Samples

1. **a)** $\dfrac{60.2}{20}$ [1]; 3.01 [1]

   b) Trace samples treated as zero   [1]
   c) Not random [1], as no sampling frame defined and not all items have an equal chance of selection. [1]
   d) Simple random sampling   [1]
   Advantage – reduction of bias [1]; disadvantage – more time consuming [1]
   e) **i)** Not as reliable as extrapolation [1]; May trend might not continue into summer. [1]
   ii) Take a sample of data from June to August.   [1]

Page 101: Measures of Central Tendency and Spread

1. **a)** $\dfrac{21.4}{10}$ [1]; 2.14 [1]

   b) $\dfrac{520}{10} - \left(\dfrac{21.4}{10}\right)^2$ [2 if fully correct; 1 for use of correct formula]

   $= 47.4204$ [1]
   c) $\sqrt{47.4204} = 6.89$ [1]

2. **a)** $\dfrac{433}{7}$ [1]; 61.9 (3 s.f.) [1]

   b) $\dfrac{7}{2} = 3.5 \therefore$ 4th value [1]; 63 [1]

3.

Number of breakdowns (x)	Frequency (f)	fx	fx^2
0	3	0	0
1	5	5	5
2	4	8	16
3	3	9	27
4	3	12	48
5	2	10	50
	20	44	146

Remember to square x before multiplying by f.

a) 44 seen [1]; $\dfrac{44}{20}$ [1] = 2.2 [1]

b) 146 seen   [1]

$\sqrt{\dfrac{146}{20} - \left(\dfrac{44}{20}\right)^2}$ [1] = 1.57 [1]

4.

Distance thrown (m)	Frequency	Cumulative frequency
$15 \leqslant d < 20$	5	5
$20 \leqslant d < 25$	10	15
$25 \leqslant d < 30$	15	30
$30 \leqslant d < 35$	12	42
$35 \leqslant d < 45$	6	48

a) $\sum f = 48$; $\frac{48}{2} = $ 24th value, in group 25 – 30 **[1]**

$\frac{m-25}{30-25} = \frac{24-15}{30-15}$ **[1]**

Median = 28 **[1]**

b) $\sum f = 48$, $\frac{48}{4} = $ 12th value, in group 20 – 25

$\frac{Q_1 - 20}{25 - 20} = \frac{12 - 5}{15 - 5}$ $Q_1 = 23.5$ **[1]**

$\sum f = 48$; $\frac{3 \times 48}{4} = $ 36th value, in group 30 – 35

$\frac{Q_3 - 30}{35 - 30} = \frac{36 - 30}{42 - 30}$ $Q_3 = 32.5$ **[1]**

IQR = 32.5 – 23.5 = 9 **[1]**

Page 102: Displaying and Interpreting Data 1 & 2

1. a) Continuous data **[1]**

b) 10 squares is a frequency of 30 so each square represents a frequency of 3.

Weight (kg)	Frequency	Cumulative frequency
$30 \leqslant w < 34$	18	18
$34 \leqslant w < 38$	30	48
$38 \leqslant w < 40$	30	78
$40 \leqslant w < 44$	36	114
$44 \leqslant w < 50$	18	132

[3 if all correct; 2 if only one error; 1 for evidence of understanding that one square = 3]

c) $\left(\left(\frac{1}{4} \times 30 \right) + 30 + \left(\left(\frac{1}{4} \right) \times 36 \right) \right) = 46.5$ **[2]**

d)

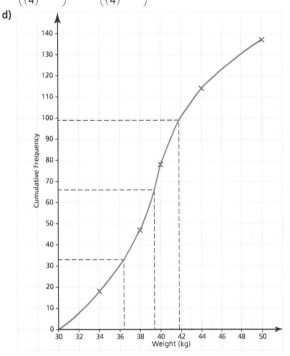

[1 for points plotted correctly; 1 for a smooth curve]

e) $Q_2 = $ answers in the range 39.0–39.4 **[1]**

$Q_1 = $ answers in the range 36.2–36.6 and $Q_3 = $ answers in the range 41.6–42.0 **[1]**

2. a)

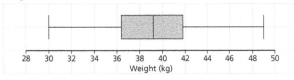

b) The young male pigs weigh more on average with a median of 49.5 kg compared to 39.2 kg **[1]**. The young male pigs have a greater spread of weight with an interquartile range of 14 kg compared to 5.4 kg **[1]**.

Page 103: Calculating Probabilities

1. a) $x = 30 - (5 + 7 + 8 + 4)$ **[1]**; $= 6$ **[1]**

b) 0 **[1]**

c) $\frac{5 + 7 + 8}{30} = \frac{20}{30}$ **[1]** $= \frac{2}{3}$ **[1]**

2. a)

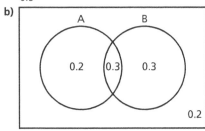

[1 for each correct pair of branches]

b) $\frac{8}{20} \times \frac{12}{19}$ or $\frac{12}{20} \times \frac{8}{19}$ **[1]**

$\frac{24}{95} + \frac{24}{95}$ **[1]** $= \frac{48}{95}$ **[1]**

3. a) $\frac{0.3}{0.5}$ **[1]**; 0.6 **[1]**

b)

[1 for 0.3 at centre; 1 for 0.2 and 0.3 correct; 1 for 0.2 outside]

c) $0.2 + 0.3 + 0.3$ **[1]** $= 0.8$ **[1]**

Page 104: Statistical Distributions

1. a) $^{10}C_3 (0.1)^3 (0.7)^7$ **[1]**; $= 0.0574$ (3 s.f.) **[1]** or $0.9872 - 0.9298$ (from tables) **[1]** $= 0.0574$ **[1]**

b) 0.9298 **[1]**

c) $1 - P(X \leqslant 5) = 1 - 0.9999$ **[1]** $= 0.0001$ **[1]**

2. $X = $ Number of tails $X \sim B(7, 0.5)$

a) $(0.5)^7 = \frac{1}{128}$ **[1]**

b) $P(X > 5) = 1 - P(X \leqslant 5)$; $1 - 0.9375$ **[1]** $= 0.0625$ **[1]**

c) $P(X < 2) = P(X \leqslant 1)$ **[1]**; 0.0625 **[1]**

3. a) 9 **[1]**   **b)** 1 **[1]**

4.

q	4	6	12
$P(Q = q)$	$\frac{1}{2}$	$\frac{1}{3}$	$\frac{1}{6}$

[2 if fully correct; 1 for one correct probability]

Page 105: Hypothesis Testing

1. a) A hypothesis test is the process by which you test a statement made about a population parameter. **[1]**

b) A critical region is the region of the probability distribution which, if the test statistic falls within it, you would reject the null hypothesis. **[1]**

c) The significance level is the probability of incorrectly rejecting the null hypothesis. **[1]**

2. $X = $ Number of seeds which do not germinate

$X \sim B(50, 0.1)$ **[1]**

$H_0: p = 0.1$ $H_1: p > 0.1$ **[1]**

$P(X \geqslant 9) = 1 - P(X \leqslant 8) = 1 - 0.9421$ **[1]** $= 0.0579$ **[1]**

0.0579 > 0.05 therefore not enough evidence to reject H_0; cannot say the company's claim is untrue. **[1]**

3. a) $P(X \leqslant 5) = 0.0233$ or $P(X \leqslant 6) = 0.0586$ **[1]**

0.0233 < 0.025 so the critical region is $X \leqslant 5$ **[1]**

$P(X \geqslant 16) = 1 - 0.9699 = 0.0301$ or $P(X \geqslant 17) = 1 - 0.9876 = 0.0124$ **[1]**; 0.0124 < 0.025 so the critical region is $X \geqslant 17$ **[1]**

b) $0.0233 + 0.0124 = 0.0357$ or 3.57% **[1]**

c) 3 lies within the critical region therefore enough evidence to reject H_0: $p \neq 0.35$ **[1]**

d) Probability of incorrectly rejecting the null hypothesis. **[1]**

4. $X = $ Number of people who prefer Crumbly Crumb biscuits **[1]**

$X \sim B(15, 0.5)$ **[1]**

$H_0: p = 0.5$ **[1]** $H_1: p > 0.5$ **[1]**

$P(X \geqslant 10) = 1 - P(X \leqslant 9)$ **[1]**; $1 - 0.8491 = 0.1509$ **[1]**

0.1509 > 0.05; Insufficient evidence to reject H_0; the company's claim cannot be supported. **[1]**

Page 106: Indefinite Integrals

1. a) $\frac{x^7}{7} - x^2 + 5x + c$ [1]

   b) $-\frac{1}{15}p^{-5} + c = -\frac{1}{15p^5} + c$ [1]

   c) $\frac{3t^2}{2} - \frac{1}{2t^2} + 3t + c$ [1]

   d) $-\frac{1}{3}x^{-3} + \frac{2}{5}x^{\frac{-5}{2}} + c$ [1]

   e) $8s^{\frac{1}{2}} + \frac{5}{2}s^2$ [1]

2. $x^5 - \frac{3}{8}x^4 - 2x^{-2} + c$ **[3 if fully correct; 1 for increase of a power by 1; 1 for two terms correct]**

3. $\frac{6}{5x^2} - \frac{2\sqrt{x}}{5x^2}$ [1]; $\frac{6}{5}x^{-2} - \frac{2}{5}x^{\frac{-3}{2}}$ [1]

   $\frac{-6}{5}x^{-1} + \frac{4}{5}x^{\frac{-1}{2}} + c$ [2]

4. a) $1 + 10\sqrt{x} + 25x$ [2]

   b) $x + \frac{20}{3}x^{\frac{3}{2}} + \frac{25x^2}{2} + c$
   [3 if fully correct; 1 for increase of a power by 1]

5. $y = 2x^2 - 2x + c$ [1]
   When $y = 5$, $x = -2$
   $5 = 2(-2)^2 - 2(-2) + c$; $c = -7$ [1]
   $y = 2x^2 - 2x - 7$ [1]

Page 106: Integration to Find Areas

1. a) $\left[\frac{-3t^4}{4} + 6t\right]_{-1}^{3}$ [1] $= -36$ [1]

   b) $\left[\frac{-q^3}{3} + \frac{5q^2}{2}\right]_{10}^{20}$ [1] $= -\frac{4750}{3}$ [1]

   c) $\left[-x^{-2} + \frac{2}{7}x^{-1}\right]_{1}^{3}$ [1] $= \frac{44}{63}$ [1]

   d) $\left[-x^{-1} - \frac{1}{2}x^{-2}\right]_{-1}^{2}$ [1] $= -\frac{9}{8}$ [1]

2. $\frac{4}{3}x^3 + \frac{3}{2}x^2 - x$ **[2 if fully correct; 1 for increase of a power by 1]**

   $\left(\frac{4}{3}(4)^3 + \frac{3}{2}(4)^2 - 4\right) - \left(\frac{4}{3}(1)^3 + \frac{3}{2}(1)^2 - 1\right)$ [1] $= \frac{207}{2}$ square units [1]

3. $\int_0^3 (x^3 - 9x)\,dx = \left[\frac{x^4}{4} - 9\frac{x^2}{2}\right]_0^3$ [1]; $\frac{-81}{4}$ or $\frac{81}{4}$ [1]

   $\frac{81}{4} \times 2$ [1] $= \frac{81}{2}$ [1]

4.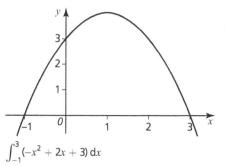

Sketch the graph.

   $\int_{-1}^{3} (-x^2 + 2x + 3)\,dx$ [1]

   $\frac{-x^3}{3} + x^2 + 3x$ [1]

   $\left(\frac{-(3)^3}{3} + (3)^2 + 3(3)\right) - \left(\frac{-(-1)^3}{3} + (-1)^2 + 3(-1)\right)$ [1] $= \frac{32}{3}$ [1]

Page 107: Introduction to Vectors

1. Unit vector is $\frac{\mathbf{i} - \mathbf{j}}{\sqrt{1^2 + 1^2}}$ [1] $= \frac{\mathbf{i} - \mathbf{j}}{\sqrt{2}}$ [1]

   So require $\frac{12(\mathbf{i} - \mathbf{j})}{\sqrt{2}}$ [1] $= 6\sqrt{2}\mathbf{i} - 6\sqrt{2}\mathbf{j}$ [1]

2. a) $|\mathbf{p}| = \sqrt{\left(\frac{\sqrt{3}}{2}\right)^2 + \left(\frac{1}{2}\right)^2}$ [1] $= 1$ [1]

   Direction $= \tan^{-1}\left(\frac{\frac{1}{2}}{\frac{\sqrt{3}}{2}}\right) = \tan^{-1}\left(\frac{1}{\sqrt{3}}\right)$ [1]

   $= 30°$ above the horizontal [1]

   b) $|\mathbf{p}| = \sqrt{4^2 + 4^2}$ [1] $= \sqrt{32} = 4\sqrt{2}$ [1]

   Direction $= \tan^{-1}\left(-\frac{4}{4}\right)$ [1]

   $= -45°$, so direction is 45° below the horizontal [1]

3. Using parallelogram law: [1]
   $R^2 = 6^2 + 10^2 - 2 \times 6 \times 10\cos60°$ [1] $\Rightarrow R = 8.72$ [1]
   Using sine rule: [1]
   $\frac{\sin\theta}{6} = \frac{\sin60}{8.72}$ [1] $\Rightarrow \theta = 36.6°$ [1]

Page 108: Vector Geometry

1. $d^2 = (2k + 1)^2 + (1 - k)^2$ [2]
   $= 5k^2 + 2k + 2$ [1]
   $= 5\left(k^2 + \frac{2k}{5} + \frac{2}{5}\right) = 5\left(\left(k + \frac{1}{5}\right)^2 + \frac{9}{25}\right)$
   $= 5\left(k + \frac{1}{5}\right)^2 + \frac{9}{5}$ [2]
   Minimum value of d^2 is $\frac{9}{5}$ [1]
   So minimum value of d is $\frac{3}{\sqrt{5}} = \frac{3\sqrt{5}}{5}$ [1]

2. $\vec{OC} = \vec{OD} + \vec{DC}$ [1]
   $= (\vec{OA} + \vec{AD}) + \vec{AB}$ [1]
   $= \begin{pmatrix} -1 \\ 4 \end{pmatrix} + \begin{pmatrix} 9 \\ -7 \end{pmatrix} + \begin{pmatrix} 4 \\ 6 \end{pmatrix}$ [1] $= \begin{pmatrix} 12 \\ 3 \end{pmatrix}$ [1]
   So coordinates are C(12, 3) [1]

3. $\vec{AB} = \begin{pmatrix} -2 \\ -5 \end{pmatrix}$ [1]; $\vec{AB} = \frac{2}{9}\vec{AC}$ [1]
   $\therefore \vec{AC} = \frac{9}{2}\begin{pmatrix} -2 \\ -5 \end{pmatrix} = \begin{pmatrix} -9 \\ -22.5 \end{pmatrix}$ [1]
   $\vec{OC} = \vec{OA} + \vec{AC} = \begin{pmatrix} 10 \\ 9 \end{pmatrix} + \begin{pmatrix} 9 \\ -22.5 \end{pmatrix}$ [1] $= \begin{pmatrix} 1 \\ -13.5 \end{pmatrix}$ [1]

Page 109: Proof

1. DAC $= 180 - 90 - x - (90 - 2x)$; Angles in triangle $= 180°$ [1]
   DAC $= x$ [1]
   DAC $=$ ACD therefore isosceles triangle [1]

2. $(4n^2 - 4n + 1) + (4n^2 + 4n + 1)$ **[2]** $= 8n^2 + 1$ $\therefore$ shown **[1]**

3. $(x + y)(x - y)$; $x^2 - xy + xy - y^2$ **[2]** $= x^2 - y^2$ $\therefore$ shown **[1]**

4. $x^2 - y^2$ $x - y = 4$
   $(y + 4)^2 - y^2$ Use simultaneous equations. [1]
   $y^2 + 8y + 16 + y^2 = 8y + 16$ [1]
   $8(y + 2)$; multiple of 8, therefore shown [1]

5. $3^2 = 9$; 9 is odd $\therefore$ shown [1]

6. $a = b \therefore a - b = 0$ [1]
   Line 4 of the proof $(a - b)(a + b) = b(a - b)$; cannot divide by 0 so proof fails here. [1]

Page 111 Quick Test

1. a) $-6\,\text{ms}^{-1}$ b) $5\,\text{ms}^{-1}$ 2. $\frac{25}{4}$ N   3. 366 N less

Page 113 Quick Test

1.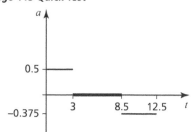

2. a)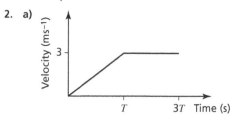

   b) $T = \frac{40}{3}$ seconds

Page 115 Quick Test

1. $t = 0.427$ seconds
2. $t = \dfrac{2\sqrt{3}}{3}$ seconds

Page 117 Quick Test

1. 21.54 m
2. $a = 7$, $b = -2$
3. 400 N

Page 119 Quick Test

1. 5.68 seconds; 55.6 ms^{-1}. Assume gravity constant, no air resistance.
2. Mass of P : Mass of Q, 2 : 1

Pages 120–123 Practice Questions

Page 120: Quantities and Units in Mechanics

1. a) By Newton 1, resistive force = 12 N since moving with constant velocity [2]
   b) By Newton 2, $F = ma$. So $(15 - 12) = 10 \times a$ [2]
   i.e. $a = 0.3\,\text{ms}^{-2}$ [1]
2. Displacement = -15 [1]
   So time $= \dfrac{\text{displacement}}{\text{velocity}} = \dfrac{-15}{-3.5} = 4.29$ seconds [2]

Page 120: Displacement–time and Velocity–time Graphs

1. a)

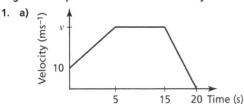

[3]

   b) Distance under graph = 850 [1]
   $\dfrac{1}{2}(10 + V)5 + 10V + \dfrac{1}{2} \times 5V = 850$ [3]
   Solve to give $V = 55\,\text{ms}^{-1}$ [1]
   c) Deceleration $= \dfrac{55}{5} = 11\,\text{ms}^{-2}$ [2]

2. a)

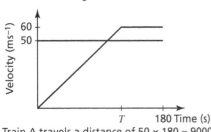

[3]

   b) Train A travels a distance of $50 \times 180 = 9000$ m [1]
   Area under graph for B must also equal 9000
   $\dfrac{1}{2} \times 60 \times T + 60(180 - T) = 9000$ [3]
   Solve to give $T = 60$ seconds [1]
   c) Initial acceleration of B is $1\,\text{ms}^{-2}$ [1]
   So A and B have the same speed after 50 seconds. [1]

Page 121: Kinematics with Constant and Variable Acceleration

1. a) $s = ut + \dfrac{1}{2}at^2$ [1] $= 2 \times 10 + \dfrac{1}{2} \times 1.5 \times 10^2 = 95$ m [1]
   b) $v^2 = u^2 + 2as$ [1]
   $v^2 = 2^2 + 2 \times 1.5 \times 47.5 \Rightarrow v = 12.1\,\text{ms}^{-1}$ [1]
2. a) $2t^2 - 9t + 4 = 0$ [1]
   $(2t - 1)(t - 4) = 0$ [1]
   $t = \dfrac{1}{2}$ or $t = 4$ [1]
   b) $s = \displaystyle\int v\,\text{d}t = \int \left(2t^2 - 9t + 4\right)\text{d}t$
   $= \dfrac{2t^3}{3} - \dfrac{9t^2}{2} + 4t + c$ [2]
   Substitute $t = 0$, $s = \dfrac{5}{2}$ [1] $\Rightarrow c = \dfrac{5}{2}$ [1]
   $s = \dfrac{2t^3}{3} - \dfrac{9t^2}{2} + 4t + \dfrac{5}{2}$ At $t = 2$, $s = -\dfrac{13}{6}$ [1]

Page 122: Forces and Newton's Laws 1

1. Using $\mathbf{a} = \dfrac{\mathbf{v} - \mathbf{u}}{t}$ [1]
   $\mathbf{a} = \dfrac{(10\mathbf{i} + 6\mathbf{j}) - (2\mathbf{i} + \mathbf{j})}{2} = \dfrac{8\mathbf{i} + 5\mathbf{j}}{2} = 4\mathbf{i} + 2.5\mathbf{j}$ [2]
   Since $|\mathbf{F}| = m|\mathbf{a}|$, $m = \dfrac{10}{\sqrt{4^2 + 2.5^2}} = 2.12$ kg [2]

2. a) Applying $F = ma$ to B:
   $R - 2g = 2a$ [2]
   $R = 1 + 2g = 20.6$ N [1]
   b) Applying $F = ma$ to A:
   $R_2 - 20.6 - 3g = 3a$ [2]
   $R_2 = 1.5 + 3g + 20.6 = 51.5$ N [1]

Page 123: Forces and Newton's Laws 2

1. a) Newton 2 to $2m$ mass: $10 - T = 2m \times 0.5$ [2]
   Newton 2 to m mass: $T = m \times 0.5$ [2]
   Add equations to give $10 = \dfrac{3m}{2}$, so $m = \dfrac{20}{3}$ kg [1]
   b) $T = \dfrac{10}{3}$ kg [1]

2. Applying $s = ut + \dfrac{1}{2}at^2$ to Joey: $s = 5t^2$ [2]
   Applying $s = ut + \dfrac{1}{2}at^2$ to Alexia: $s = 2(t - 5) + 5(t - 5)^2$ for $t \geqslant 5$ [2]
   Solving simultaneously: $5t^2 = 2(t - 5) + 5(t - 5)^2$
   Solving gives $t = \dfrac{115}{48}$ [2], so $s = 5 \times \left(\dfrac{115}{48}\right)^2 = 28.7$ m [1]

Pages 124–132 Review Questions

Page 124: Populations and Samples

1. a) A population is a set of items which are of interest. [1]
   b) A sample is a selection of items from the population. [1]
2. a) Number all members of the club 1–280 [1]. Use a random number generator to select 40 random numbers between 1 and 280 [1]. Survey the member which corresponds to the number generated. [1]
   b)

	Male	Female
Under 30	$\dfrac{80}{280} \times 40 = 11$	$\dfrac{60}{280} \times 40 = 9$
Over 30	$\dfrac{90}{280} \times 40 = 13$	$\dfrac{50}{280} \times 40 = 7$

[3 if all correct; 1 for evidence of use of $\dfrac{x}{280} \times 40$; 1 for 11, 9, 13 and 7; 1 for at least one correct]

Use a simple random sampling method to select the right number of members from each group. [1]

> You may have to round your answers in the table.

3. Quota sampling [1]; carrying out market research on a high street (or any other acceptable answer) [1]
4. a) $\dfrac{101.2}{6}$ [1] = 16.9 [1]
   b) $\dfrac{465.7}{24}$ [1] = 19.4 [1]
   Second estimate is more reliable as it uses more data. [1]

Page 125: Measures of Central Tendency and Spread

1.

Time (x)	Frequency (f)	fx	fx^2
20	3	60	1200
21	17	357	7497
22	29	638	14 036
23	34	782	17 986
24	26	624	14 976
25	10	250	6250
	119	2711	61 945

   a) $\dfrac{2711}{119}$ [1] = 22.8 (3 s.f.) [1]
   b) 61 945 seen [1]
   $\sqrt{\dfrac{61\,945}{119} - \left(\dfrac{2711}{119}\right)^2}$ [1] = 1.24 [1]

2. a) $71 \times \dfrac{40}{100} = 28.4$th value, in group 21–30
   $\dfrac{P_{40} - 20.5}{30.5 - 20.5} = \dfrac{28.4 - 26}{46 - 26}$ [1]; 21.7 [1]
   b) $71 \times \dfrac{80}{100} = 56.8$th value, in group 31–40
   $\dfrac{P_{80} - 30.5}{40.5 - 30.5} = \dfrac{56.8 - 46}{65 - 46}$ [1]; 36.2 [1]
   c) $36.2 - 21.7 = 14.5$ [1]

3.

Time taken (min)	Midpoint	Frequency	fx	fx^2
20–29	24.5	8	196	4802
30–39	34.5	12	414	14 283
40–49	44.5	35	1557.5	69 308.75
50–59	54.5	18	981	53 464.5
60–69	64.5	9	580.5	37 442.25
		82	3729	179 300.5

a) $\frac{3729}{82}$ **[1]** = 45.5 minutes **[1]**

b) 179 300.5 seen   **[1]**

$\sqrt{\frac{179\,300.5}{82} - \left(\frac{3729}{82}\right)^2}$ **[1]** = 10.9 **[1]**

c) 45.5 + 10.9 = 56.4   Work out what fraction of the group is greater than 56.4 and multiply this by the frequency.   **[1]**

$\frac{59.5 - 56.4}{10} \times 18 = 5.58$ **[1]**

5.58 + 9 = 14.58, therefore 14 runners   **[1]**

Page 126: Displaying and Interpreting Data 1 & 2

1. a) $\frac{30}{4} = 7.5 \therefore$ 8th value or $\frac{3 \times 30}{4} = 22.5 \therefore$ 23rd value   **[1]**

9.5 **[1]** and 10.9 **[1]**   Order the data from smallest to biggest.

b) Interquartile range = 10.9 – 9.5 = 1.4
9.5 – (1.5 × 1.4) = 7.4
10.9 + (1.5 × 1.4) = 13   **[1]**
No outliers   **[1]**

c)

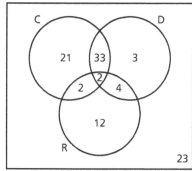

Daily mean temperature (°C)   **[3]**

2. a) Positive   **[1]**
b) Amount the mean wind speed increases per one knot increase in the maximum gust of wind.   **[1]**
c) Data points indicate a linear relationship (close to straight line).   **[1]**

Page 127: Calculating Probabilities

1.

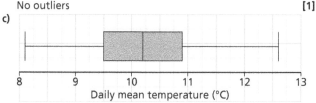

a) [1 for centre correct; 1 for 2, 4, 33 correct; 1 for 12, 21, 3 correct; 1 for 23 correct]

b) $\frac{23}{100}$ **[1]**   c) $\frac{21}{100}$ **[1]**   d) $\frac{6}{42}$ **[1]** = $\frac{1}{7}$ **[1]**

e) No – some people own both a cat and a dog   **[1]**

f) P(cat) × P(rabbit) = $\frac{58}{100} \times \frac{20}{100}$ = 0.116   **[1]**

P(cat and rabbit) = 0.04
0.04 ≠ 0.116, therefore not independent.   **[1]**

If A is independent of B, then P(A) × P(B) = P(A and B).

2. a)

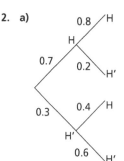

[1 for each pair of correct branches]

b) 0.7 × 0.2 **or** 0.3 × 0.4   **[1]**
0.14 + 0.12 **[1]** = 0.26 **[1]**

Page 128: Statistical Distributions

1. X = number of patients who listen to the hospital radio
a) X follows a binomial distribution **[1]**; $X \sim B(20, 0.4)$ **[1]**
b) $P(X \geq 5) = 1 - P(X \leq 4)$; 1 – 0.0510 **[1]** = 0.949 **[1]**

2. a)

X	1	2	3	4
$P(X = x)$	$\frac{1}{4}$	$\frac{1}{4}$	$\frac{1}{4}$	$\frac{1}{4}$

[1 for table with equal probabilities; 1 for probabilities = $\frac{1}{4}$]

b) Discrete uniform   **[1]**

3. a) X = Number of blue mice **[1]**; $X \sim B(25, 0.3)$ **[1]**
$P(X \leq 10) = 0.9022$ **[1]**
b) $P(X \geq 15) = 1 - P(X \leq 14)$ **[1]**
1 – 0.9982 **[1]** = 0.0018 **[1]**

4. a) All counters equally likely to be selected   **[1]**
b) i) $\frac{1}{12}$ **[1]**   ii) $\frac{5}{12}$ **[1]**   iii) $\frac{5}{12}$ **[1]**

5. a) $^{20}C_4(0.8)^4 (0.2)^{16}$ **[1]**; 1.3×10^{-8} **[1]**
b) $Y \sim B(20, 0.2)$
$P(X \geq 14) = P(Y \leq 6)$ **[1]**; 0.9133 **[1]**

Page 129: Hypothesis Testing

1. a) $B(40, p)$ **[1]**
b) $H_0: p = 0.25$ **[1]**; $H_1: p > 0.25$ **[1]**
X = No. of 3s recorded
$X \sim B(40, 0.25)$ **[1]**
$P(X \geq 15) = 1 - P(X \leq 14) = 1 - 0.9456$ **[1]** = 0.0544 **[1]**
0.0544 > 0.05 **[1]**
Not enough evidence to reject the null hypothesis.
No evidence to show that the spinner is biased.   **[1]**

2. a) $P(X \geq 8) = 1 - 0.8909 = 0.1091$ and $P(X \geq 9) = 1 - 0.9532 = 0.0468$ **[1]**   0.0468 < 0.05 **[1]**
Critical region is $X \geq 9$ **[1]**
b) $P(X \geq 12) = 1 - 0.9558 = 0.0442$ and
$P(X \geq 13) = 1 - 0.9825 = 0.0175$
$P(X \leq 2) = 0.0090$ and $P(X \leq 3) = 0.0332$ **[1]**
0.0332 > 0.025 and 0.0175 < 0.025   **[1]**

This is a two-tailed test.

Critical regions are $X \geq 13$ and $X \leq 2$ **[1]**
c) $X \sim B(25, 0.8)$ $Y \sim B(25, 0.2)$ **[1]**
$P(Y \leq 1) = 0.0274$; 0.0274 < 0.05   **[1]**
Critical region is $X \geq 24$ **[1]**

3. $H_0: p = 0.3$ **[1]**; $H_1: p > 0.3$ **[1]**
$X \sim B(20, 0.3)$ **[1]**
$P(X \geq 11) = 1 - P(X \leq 10)$ **[1]** = 0.0171 **[1]**
0.0171 < 0.10 **[1]** Evidence to reject the null hypothesis; practice has improved chance of success. **[1]**

Page 130: Quantities and Units in Mechanics

1. Acceleration = $\frac{\text{force}}{\text{mass}} = \frac{12}{15} = 0.6\,\text{ms}^{-2}$ **[2]**

Increase in velocity = 0.6 × 5 = 3 ms⁻¹   **[1]**

2. A: $0.45 = \frac{X + 5}{T} \Rightarrow T = \frac{X + 5}{0.45}$ **[2]**

B: $0.3 = \frac{X - 10}{T} \Rightarrow T = \frac{X - 10}{0.3}$ **[2]**

So $\frac{X + 5}{0.45} = \frac{X - 10}{0.3}$ **[1]**
Solve to give $X = 40$ **[1]**

3. Change in velocity = acceleration × time **[1]**

$34 - u = 9.8 \times 3.2$ **[1]** So $u = 34 - 9.8 \times 3.2 = 2.64\,\text{ms}^{-1}$ **[1]**

Page 130: Displacement–time and Velocity–time Graphs

1. a)
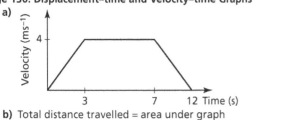
[3]

b) Total distance travelled = area under graph **[1]**

$= \frac{1}{2}(12 + 4)4 = 32\,\text{m}$ **[1]**

Let T be the time at which cyclist completes half his journey.

Then $\frac{1}{2} \times 4 \times 3 + 4(T - 3) = 16$ **[1]**   So $T = 5.5$ s **[1]**

2. a)
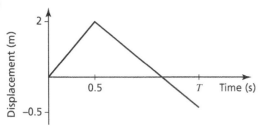
[3]

b) Deceleration $= \dfrac{\text{change in velocity}}{\text{time}}$

$3.5 = \dfrac{2.5}{(T - 0.5)}$ **[2]**

$T = \dfrac{17}{14} = 1.21$ s **[1]**

So average speed $= \dfrac{\text{total distance}}{\text{total time}}$ **[1]**

$= \dfrac{\left(\frac{9}{2}\right)}{\left(\frac{17}{14}\right)} = \dfrac{63}{17} = 3.71\,\text{ms}^{-1}$ **[1]**

Page 131: Kinematics with Constant and Variable Acceleration

1. $s = vt - \frac{1}{2}at^2$ **[1]** $26 = 8t - \frac{1}{2} \times 0.75t^2$ **[1]**

$3t^2 - 64t + 208 = 0$ **[1]** $t = 4$ or $t = 17.33$

$t = 4$ since this is the first time the displacement is 26 m. **[1]**

2. a) $v = \dfrac{ds}{dt} = t^2 - 3t + \dfrac{5}{2}$ **[2]**

Attempt to complete the square: $v = \left(t - \dfrac{3}{2}\right)^2 + \dfrac{1}{4}$ **[1]**

$v > 0$ for all t, hence result. **[1]**

b) $a = \dfrac{dv}{dt} = 2t - 3$ **[1]**

$2t - 3 < 0 \Rightarrow 0 < t < \dfrac{3}{2}$ **[2]**

3. $v = \int a\,dt = \int (2t)\,dt = t^2 + c$ **[2]**

At $t = 3$, $v = -3$

$-3 = 9 + c \Rightarrow c = -12$ **[1]**

So $v = t^2 - 12$

$s = \int (t^2 - 12)\,dt$ **[1]** $= \dfrac{t^3}{3} - 12t + c$ **[1]**

$t = 0$, $s = 0 \Rightarrow c = 0$

$s = \dfrac{t^3}{3} - 12t = t\left(\dfrac{t^2}{3} - 12\right)$ **[1]**

Returns to starting point when $s = 0$, i.e. $t = 6$ s **[1]**

Page 131: Forces and Newton's Laws 1

1. Using $\mathbf{s} = \left(\dfrac{\mathbf{u} + \mathbf{v}}{2}\right)t$ **[1]**

$\mathbf{s} = \left(\dfrac{(-2\mathbf{i} - 10\mathbf{j}) + (-\mathbf{i} + 8\mathbf{j})}{10}\right)t$ **[1]**

$\mathbf{s} = \dfrac{(-3\mathbf{j} - 2\mathbf{j})}{2} \times 10 = -15\mathbf{i} - 10\mathbf{j}$ **[1]**

$|\mathbf{s}| = \sqrt{15^2 + 10^2} = \sqrt{325} = 5\sqrt{13}$ **[2]**

---

2. Applying Newton 2 downwards gives: $35g - R = 35a$ **[2]**

So $R = 35g - 35 \times 2 = 273\,\text{N}$, which is the reaction force on the child. **[1]**

By Newton 3, the force the child exerts on the lift is therefore also 273 N. **[1]**

3. Newton 1 implies the resistive force is 30 N **[1]**

Applying Newton 2 when car is accelerating: $50 - 30 = 150a$ **[2]**

$\Rightarrow a = \dfrac{2}{15}\,\text{ms}^{-2}$ **[1]**

Page 132: Forces and Newton's Laws 2

1. a) Applying $v = u + at$ upwards: $-\dfrac{u}{2} = u - gT$ **[2]**

So $T = \dfrac{3u}{2g}$ **[1]**

b) Applying $v^2 = u^2 + 2as$: $\left(-\dfrac{u}{2}\right)^2 = u^2 - 2gs$ **[2]**

So $s = \dfrac{3u^2}{8g}$ **[1]**

2. Applying $s = ut + \frac{1}{2}at^2$ upwards: $-100 = 15t - 4.9t^2$ **[2]**

$4.9t^2 - 15t - 100 = 0$

$t = \dfrac{15 \pm \sqrt{15^2 - 4 \times 4.9 \times (-100)}}{9.8}$ **[2]**

$t = 6.30$ seconds (only positive solution) **[1]**

3. a) Newton 2 to M kg mass: $\quad Mg - T = 0.75M$ **[2]**

Newton 2 to 5 kg mass: $\quad T - 5g = 5 \times 0.75$ **[2]**

Adding equations: $\quad Mg - 5g = 3.75 + 0.75M$

Solving: $M = \dfrac{3.75 + 5g}{g - 0.75} = 5.83\,\text{kg}$ **[2]**

b) When M mass hits floor, velocity of 5 kg mass is given by:

$v^2 = u^2 + 2as = 0 + 2 \times \dfrac{3}{4} \times 2 = 3$ **[1]**

Now 5 kg moves freely under gravity, so $v^2 = u^2 + 2as$

$\Rightarrow 0 = 3 + 2 \times (-9.8) \times s$ **[1]**; $\Rightarrow s = 0.153\,\text{m}$ **[1]**

When M hits the floor, the 5 kg mass is at 4 m as it has travelled 2 m upwards having started from 2 m above the floor. So height above floor reached is $4 + 0.153 = 4.15\,\text{m}$ **[1]**

Pages 133–142 Mixed Questions

Pure Mathematics

1. $a + b = ab \qquad a = b + 2$

$b + 2 + b = b(b + 2)$ **[1]**

$2b + 2 = b^2 + 2b \therefore b^2 + 2 = 0$ **[1]**

$b = \pm\sqrt{2} \therefore a = 2 \pm \sqrt{2}$ not an integer **[1]**

2. a) $\left[\dfrac{3x^2}{2} + 2x\right]_0^4$ **[1]** $= 32$ **[1]**

b) $\left[24t - 2t^3\right]_{-1}^1$ **[1]** $= 44$ **[1]**

c) $\left[1.6s^2 + 14.6s\right]_{4.7}^{5.1}$ **[1]** $= 12.112$ **[1]**

d) $\left[\dfrac{-2}{3}x^{-3} - \dfrac{3}{5}x^{-1}\right]_1^2$ **[1]** $= \dfrac{53}{60}$ **[1]**

3. a) $2x + 3 + 2x + 3 + 3x - 4 + 3x - 4 = 10x - 2$ **[1]**

b) $(2x + 3)(3x - 4) = 6x^2 + x - 12$ **[1]**

c) $(2x + 3)(3x - 4) = 6x^2 + x - 12$, so after removing 3 cm² the area is $6x^2 + x - 15$ **[1]** $= (2x - 3)(3x + 5)$ **[1]**

4. a) BC: $d_{BC} = \sqrt{(10 - 8)^2 + (0 - 4)^2} = \sqrt{20} = 2\sqrt{5}$ units **[1]**

b) $m_{BC} = \dfrac{0 - 4}{10 - 8} = -2$ **[1]**

Gradient from A to D is $\dfrac{1}{2}$ **[1]**

Equation of BC: $y - 4 = -2(x - 8) \Rightarrow y = -2x + 20$ **[1]**

Equation of line AD:

$y - (-3) = \dfrac{1}{2}(x - (-1)) \Rightarrow y = \dfrac{1}{2}x - \dfrac{5}{2}$ **[1]**

c) Point D is where line AD and line BC intersect:

$-2x + 20 = \dfrac{1}{2}x - \dfrac{5}{2}$

[1 mark for simultaneous equations and attempt to solve]

$-4x + 40 = x - 5 \Rightarrow x = 9$

$y = -2x + 20 \therefore y = (-2 \times 9) + 20 \Rightarrow y = 2$

Point D (9, 2) **[1]**

$d_{AD} = \sqrt{(9 - (-1))^2 + (2 - (-3))^2} = \sqrt{125} = 5\sqrt{5}$ **[1]**

d) Area of triangle $= \frac{1}{2}b \times h = \frac{1}{2} \times \left(2\sqrt{5}\right) \times \left(5\sqrt{5}\right) = 25$ units² **[1]**

5. $V = \pi r^2 h$ $\qquad\qquad\qquad\qquad \pi r^2 h = 350$ **[1]**

$SA = 2\pi r^2 + 2\pi rh$ **[1]** $h = \dfrac{350}{\pi r^2}$

$SA = 2\pi r^2 + 2\pi r \times \dfrac{350}{\pi r^2} = 2\pi r^2 + \dfrac{700}{r}$ **[2]**

$\dfrac{d(SA)}{dr} = 4\pi r - \dfrac{700}{r^2}$ **[1]**

$4\pi r - \dfrac{700}{r^2} = 0, r^3 = \dfrac{700}{4\pi}$ **[1]**

$r = 3.82$ cm (3 s.f.), $h = 7.63$ cm **[1]**

6. Using the discriminant, $b^2 - 4ac > 0$,
$(8k)^2 - (4 \times 3 \times (-4k)) = 64k^2 + 48k > 0$ **[1]**
Finding the x-intercepts:

$0 = 64x^2 + 48x = 16x(4x + 3), x = 0$ and $x = -\dfrac{3}{4}$ **[1]**

Sketching the graph of $y = 64x^2 + 48x$ **[1]** and identifying
the values of x for which $y > 0$, $x < -\dfrac{3}{4}$ or $x > 0$, hence

$\left\{k : k < -\dfrac{3}{4}\right\} \cup \{k : k > 0\}$. **[1]**

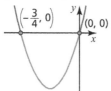

7. a)

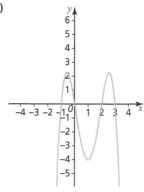

[1 mark for each x-intercept; general shape must be correct to earn any marks]

b) $y = f(x)$ intersects the x-axis at $(-1, 0)$, $(0, 0)$, $(2, 0)$ and $(3, 0)$,
so $y = f(x + a)$ would be a translation of $-1, -2, -4$ or -5 in
the x-direction. Thus $a = 1, a = 2, a = 4$ or $a = 5$.
[1 mark for each value of a]

8. $\dfrac{30x^4 y^{\frac{3}{2}}}{15x^2 y^{\frac{7}{2}}} = 2x^2 \times y^{\frac{3}{2} - \frac{7}{2}} = 2x^2 y^{-2}$ **[1]**

9. $\ln(5x) = 0 \Rightarrow 5x = 1$ **[1]**; $x = \dfrac{1}{5}$ **[1]**

Hence, solution to inequality is $0 < x < \dfrac{1}{5}$ **[2]**

10. $\mathbf{a} - 2\mathbf{b} = (4\mathbf{i} + 5\mathbf{j}) - 2(\mathbf{i} + 3\mathbf{j}) = 2\mathbf{i} - \mathbf{j}$ **[1]**
Required vector is of the form $2k\mathbf{i} - k\mathbf{j}$ **[1]**
$(2k)^2 + (-k)^2 = 100$ **[1]**
$5k^2 = 100$, so $k = \sqrt{20} = 2\sqrt{5}$ **[2]**
Vector is therefore $4\sqrt{5}\mathbf{i} - 2\sqrt{5}\mathbf{j}$ **[1]**

11. Writing a set of simultaneous equations:
$h + 4r = 26$ and $2\pi r^2 + 2\pi rh = 112\pi$ **[1]**
Rearranging and substituting: $2\pi r^2 + 2\pi r(26 - 4r) = 112\pi$
$2\pi r^2 + 52\pi r - 8\pi r^2 = 112\pi$
$-6r^2 + 52r - 112 = 0$ **[1]**
$(r - 4)(3r - 14) = 0$

$r - 4 = 0, r = 4$ or $3r - 14 = 0, r = \dfrac{14}{3}$ **[1]**

Substitute $r = 4, h + (4 \times 4) = 26, h = 10$

$r = \dfrac{14}{3}, h + \left(4 \times \dfrac{14}{3}\right) = 26, h = \dfrac{22}{3}$ **[1]**

Finding the volume of both cases: $V = \pi r^2 h$
$= \pi \times (4^2) \times 10 = 160\pi$ cm³ **[1]**

$V = \pi r^2 h = \pi \times \left(\dfrac{14}{3}\right)^2 \times 10 = \dfrac{1960}{9}\pi \approx 218\pi$ cm³ **[1]**

12. a) $y = \dfrac{9}{x}$ $\quad$ Given $y = \dfrac{a}{x}$, look at the y-value when $x = 1$ to find the value of a. **[1]**

b) $(x + 3)(x + 1)(2x - 3) = 0$
$(x^2 + 4x + 3)(2x - 3) = 0$ **[1]**
$2x^3 + 5x^2 - 6x - 9 = 0$ **[1]**
$2x^3 + 5x^2 - 6x = 9$

$2x^2 + 5x - 6 = \dfrac{9}{x}$ **[1]**

c) From part **b)**, the points of intersection are $x = -3$, $x = -1$
and $x = \dfrac{3}{2}$. Substituting these values into either equation to
find the corresponding y-values:

$y = \dfrac{9}{-3} = -3$, $y = \dfrac{9}{-1} = -9$, $y = \dfrac{9}{\left(\frac{3}{2}\right)} = 6$ **[1]**

$(-3, 3)$, $(-1, -9)$ and $\left(\dfrac{3}{2}, 6\right)$ **[1]**

13. $(2\log_2 x - 3)(2\log_2 x - 7) = 0$ **[2]**

$2\log_2 x - 3 = 0 \Rightarrow \log_2 x = \dfrac{3}{2} \Rightarrow x = 2^{\frac{3}{2}}$ **[2]**

$\Rightarrow x = \left(\sqrt{2}\right)^3 = 2\sqrt{2}$ **[1]**

$2\log_2 x - 7 = 0 \Rightarrow \log_2 x = \dfrac{7}{2} \Rightarrow x = 2^{\frac{7}{2}}$ **[2]**

$\Rightarrow x = \left(\sqrt{2}\right)^7 = 8\sqrt{2}$ **[1]**

14.

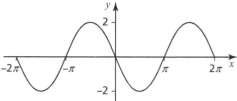

[1 for correct shape; 1 for 2 and -2 maximum and minimum; 1 for $-2\pi, -\pi, 0, \pi$ and 2π indicated as crossing points on x-axis]

15. a) $\sqrt{1 - \sin^2\theta}$ **[1]**; $\cos\theta = c$ **[1]**

b) $\dfrac{1 - \cos^2\theta}{\cos^2\theta} = \dfrac{\sin^2\theta}{\cos^2\theta}$ **[1]**; $\tan^2\theta = t^2$ **[1]**

16. a) $\dfrac{-1}{p} + c$ **[1]**

b) $2t + \dfrac{1}{t} + c$ **[1]**

c) $\dfrac{-2}{x} + \dfrac{4}{\sqrt{x}} + c$ **[1]**

d) $\dfrac{1}{3}x^3 - \dfrac{1}{2}x^2 - 12x + c$ **[1]**

17. Let w be the width, then the length is $l = 20 - 2w$
The area is $w \times l = 40$

$w(20 - 2w) = 40$ $\quad$ Substitute $l = 20 - 2w$ **[1]**

$-2w^2 + 20w = 40$

$w^2 - 10w = -20$ $\quad$ Divide each term by -2.

$(w - 5)^2 = -20 + 25$ $\quad$ Complete the square.

$(w - 5)^2 = 5$ **[1]**

$w - 5 = \sqrt{5}$, so $w = 5 \pm \sqrt{5}$ **[1]**

$w = 5 + \sqrt{5} \approx 7.24$ m

When width is 7.24 m, length is $20 - (2 \times 7.24) = 5.52$ m **[1]**
Or $w = 5 - \sqrt{5} \approx 2.76$ m

When width is 2.76 m, length is $20 - (2 \times 2.76) = 14.48$ m **[1]**

18. If $f(-2) = 0$, then $(x + 2)$ is a factor of $f(x)$. **[1]**

$$\begin{array}{r} 6x^2 - 7x - 3 \\ x + 2 \overline{\smash{\big)}\, 6x^3 + 5x^2 - 17x - 6} \\ \underline{-\ (6x^3 + 12x^2)} \\ -7x^2 - 17x \\ \underline{-\ (-7x^2 - 14x)} \\ -3x - 6 \\ \underline{-\ (-3x - 6)} \\ 0 \end{array}$$

[1 for attempting division; 1 for quotient correct]

Two numbers whose product is $6 \times -3 = -18$ and add to -7 are
2 and -9.

$6x^2 - 7x - 3 = 6x^2 + 2x - 9x - 3$ $\quad$ Factorise $6x^2 - 7x - 3$
$= 2x(3x + 1) - 3(3x + 1)$
$= (3x + 1)(2x - 3)$ **[1 for factorising]**
$6x^3 + 5x^2 - 17x - 6 = (3x + 1)(2x - 3)(x + 2)$ **[1 for correct answer]**

19. a) Using Pythagoras' Theorem: $AC^2 + BC^2 = AB^2$ and the distance

formula $d = \sqrt{(x_2 - x_1)^2 + (y_2 - y_1)^2}$

$AC^2 = (c - (-6))^2 + (8 - 14)^2 = c^2 + 12c + 72$ **[1]**

$BC^2 = (c - 10)^2 + (8 - 2)^2 = c^2 - 20c + 136$ **[1]**

$AB^2 = (10 - (-6))^2 + (2 - 14)^2 = 400$

$c^2 + 12c + 72 + c^2 - 20c + 136 = 400$ **[1]**

$2(c^2 - 4c + 104) = 400$

$c^2 - 4c + 104 = 200$, so $c^2 - 4c = 96$

$(c - 2)^2 = 100$ Complete the square. **[1]**

$c = 12$ or $c = -8$; Given that $c < 0$, $c = -8$ **[1]**

b) AB is the diameter of the circle because angle ACB is 90°
(because the angle in a semi-circle is a right angle). **[1]**

The centre of the circle is the midpoint of AB:

$\left(\dfrac{x_1 + x_2}{2}, \dfrac{y_1 + y_2}{2}\right) = \left(\dfrac{-6 + 10}{2}, \dfrac{14 + 2}{2}\right) = (2, 8)$ **[1]**

The radius of the circle is half the distance of AB; from above
$AB^2 = 400$

$AB = \sqrt{400} = 20$, so $r = 10$ **[1]**

The equation of the circle is $(x - 2)^2 + (y - 8)^2 = 100$ **[1]**

20. $2a + 3b = \begin{pmatrix} 2\lambda \\ 6 \end{pmatrix} + \begin{pmatrix} 3 \\ -9 \end{pmatrix} = \begin{pmatrix} 2\lambda + 3 \\ -3 \end{pmatrix}$ **[2]**

$(2\lambda + 3)^2 + (-3)^2 = 34$ **[1]**, so $(2\lambda + 3)^2 = 25$ **[1]**

$2\lambda + 3 = 5$ or $2\lambda + 3 = -5$ **[2]**

$\lambda = 1$ or $\lambda = -4$ **[2]**

21. a) and b)

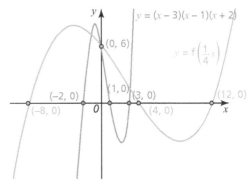

[1 for x-intercepts; 1 for y-intercept; 1 for general shape]

22. a) $\dfrac{h^3 + 3}{2h} = \dfrac{h^3}{2h} + \dfrac{3}{2h}$ **[1]**; $\dfrac{1}{2}h^2 + \dfrac{3}{2h}$ **[1]**

b) $\dfrac{1}{2}h^2 + \dfrac{3}{2}h^{-1}$ **[1]**; $h - \dfrac{3}{2}h^{-2}$ **[2]**

c) $1 + 3h^{-3}$ **[2]**

23. $3n + 3n + 3 + 3n + 6$ **[1]**

$9n + 9 = 3(3n + 3)$ **[1]** ∴ a multiple of 3 **[1]**

24. $N = N_0 e^{-\lambda t} \Rightarrow \ln N = \ln N_0 - \lambda t$ **[1]**

$\ln 200 = \ln N_0 - 5\lambda$ and $\ln 180 = \ln N_0 - 7\lambda$ **[1]**

Solving simultaneously: **[1]**

$\lambda = 0.0527$ **[1]**; $N_0 = 260$ **[1]**

25. $\sin 5\theta = 0$, $180 \leqslant \theta \leqslant 540$, ∴ $900 \leqslant 5\theta \leqslant 2700$

$\sin^{-1} 0 = 0$ **[1]**

2nd solution: $180 - 0 = 180$

Further solutions: $0 + 360 = 360$ and $180 + 360 = 540$, etc.

$5\theta = 0, 180, 360, 540, 720, 900, 1080, 1260, 1440, 1620,$
$1800, 1980, 2160, 2340, 2520, 2700$ **[1]**

$\theta = 180, 216, 252, 288, 324, 360, 396, 432, 468, 504, 540$ **[1]**

26. $y = \dfrac{7x^4}{4} + \dfrac{5x^3}{3} - \dfrac{7x^2}{2} + x + c$ **[3 if all correct; 1 for two terms correct; 1 for increasing power correctly for one term]**

Statistics and Mechanics

1. a) X = Number of people with virus

$X \sim B(50, 0.05)$ **[1]**

$P(X \geqslant 5) = 0.1036$ and $P(X \geqslant 6) = 0.0378$ **[2]**

$0.0378 < 0.05$ **[1]**; Critical region is $X \geqslant 6$ **[1]**

b) $10 > 6$ **[1]**; 10 lies within the critical region; enough evidence to suggest the rate of illness in call centre is higher than in city. **[1]**

2. a)

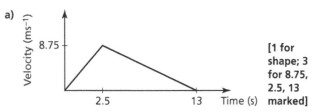

[1 for shape; 3 for 8.75, 2.5, 13 marked]

b) $\dfrac{1}{2} \times 13 \times 8.75$ **[1]** $= 56.9$ m **[1]**

3. a) $\dfrac{32}{31}$ **[1]** $= 1.03$ **[1]**

b) $\sqrt{\dfrac{78}{31} - \left(\dfrac{32}{31}\right)^2} = 1.20$ **[2]**

c) The day when there were six faults appears unusual and
will have made the mean number of faults higher **[1]**,
therefore this data might not be reliable for making
conclusions about the reliability of the production line. **[1]**

4. Acceleration $= \dfrac{\text{velocity}}{\text{time}} = -\dfrac{10}{5} = -2$ ms^{-2} **[2]**

Newton 2 implies $F = ma$, so $F = 5 \times (-2) = -10$ N **[2]**

So magnitude of force $= 10$ N **[1]**

5. $\mathbf{F_1} + \mathbf{F_2} + \mathbf{F_3} = 0$ **[1]**

$\Rightarrow (2p - q + 5q - 1)\mathbf{i} + (q + p + 2p)\mathbf{j} = 0$

$2p + 4q - 1 = 0$ **[1]**

$q + 3p = 0$ **[1]**

$q = -3p \Rightarrow 2p + 4(-3p) - 1 = 0$

So $p = -\dfrac{1}{10}$ and $q = \dfrac{3}{10}$ **[2]**

6. Whilst the scatter diagram suggests positive correlation,
this does not suggest causation **[1]**. There might be another
factor which explains the relationship, e.g. people spend
more during the summer holidays **[1]**.

7. a) $3k = 1 - (0.1 + 0.3 + 0.3)$ **[1]**; $3k = 0.3$, so $k = 0.1$ **[1]**

b) 0.6 **[1]**

8. a) 0.9619 **[1]**   **b)** 0 **[1]**   **c)** 0.9950 **[1]**

9. a) $s = ut + \dfrac{1}{2}at^2$ **[1]**

$42.5 = 5u + \dfrac{1}{2}a \times 5^2 \Rightarrow 42.5 = 5u + 12.5a$ or $85 = 10u + 25a$ **[1]**

$160 = 10u + \dfrac{1}{2}a \times 10^2 \Rightarrow 160 = 10u + 50a$ **[1]**

Solve simultaneously to give: $u = 1$ ms^{-1} **[1]**

b) $a = 3$ ms^{-2} **[1]**

10. a) i) Newton 2   6 kg mass: $T - 15 = 6a$ **[2]**

Newton 2   3 kg mass: $3g - T = 3a$ **[2]**

Add to give: $3g - 15 = 9a$

So $a = \dfrac{3g - 15}{9} = 1.6$ ms^{-2} **[1]**

ii) $T = 6a + 15 = 24.6$ N **[1]**

b) After 2 seconds, $v = u + at \Rightarrow 1.6 \times 2 = 3.2$ ms^{-1} **[1]**

Newton 6 kg mass: $-15 = 6a \Rightarrow a = -\dfrac{5}{2}$ ms^{-2} **[2]**

Applying $v^2 = u^2 + 2as$: $0 = 3.2^2 + 2 \times \left(-\dfrac{5}{2}\right)s$ **[1]**

Solving to give $s = 2.05$ m **[1]**

11. a) Advantage: Gives a complete picture of the population **[1]**

Disadvantage: Time-consuming and expensive **[1]**

b) Advantage: Less data to process or cheaper / less time-
consuming **[1]**

Disadvantage: Data may not fully represent the
population / possible introduction of bias **[1]**

12. Correct cumulative frequencies: 5, 15, 41, 49, 50 **[1]**

Median $= 96.75$ **[1]**; Lower quartile $= 94.25$ **[1]**;
Upper quartile $= 99.5$ **[1]**

13. 26 five-by-five squares in total; 104 people = 26 squares **[1]**

1 square = 4 people

14–20 is 18 squares, $18 \times 4 = 72$ **[1]**

$\dfrac{72}{104} = \dfrac{9}{13}$ **[1]**

14. $H_0: p = 0.15$ **[1]** $H_1: p > 0.15$ **[1]**

X = Number of patients with complications

$X \sim B(30, 0.15)$ **[1]**

$P(X \geqslant 10) = 1 - P(X \leqslant 9) = 1 - 0.9903$ **[1]** $= 0.0097$ **[1]**

$0.0097 < 0.01$ **[1]**

Enough evidence to reject the null hypothesis. Evidence to
suggest this hospital has more complications post-surgery. **[1]**

Glossary and Index

Collins

Edexcel A-Level
Maths
Year 1 & AS
Workbook

Phil Duxbury, Rebecca Evans
and Leisa Bovey

Revision Tips

Rethink Revision

Have you ever taken part in a quiz and thought *'I know this!'* but, despite frantically racking your brain, you just couldn't come up with the answer?

It's very frustrating when this happens but, in a fun situation, it doesn't really matter. However, in your A-level exams, it will be essential that you can recall the relevant information quickly when you need to.

Most students think that revision is about making sure you *know* stuff. Of course, this is important, but it is also about becoming confident that you can **retain** that *stuff* over time and **recall** it quickly when needed.

Revision That Really Works

Experts have discovered that there are two techniques that help with all of these things and consistently produce better results in exams compared to other revision techniques.

Applying these techniques to your A-level revision will ensure you get better results in your exams and will have all the relevant knowledge at your fingertips when you start studying for further qualifications or begin work.

It really isn't rocket science either – you simply need to:

- **test yourself** on each topic as many times as possible
- **leave a gap** between the test sessions.

Three Essential Revision Tips

Use Your Time Wisely

- Allow yourself plenty of time.
- Try to start revising at least six months before your exams – it's more effective and less stressful.
- Your revision time is precious so use it wisely – using the techniques described on this page will ensure you revise effectively and efficiently and get the best results.
- Don't waste time re-reading the same information over and over again – it's time-consuming and not effective!

Make a Plan

- Identify all the topics you need to revise (this Complete Revision & Practice book will help you).
- Plan at least five sessions for each topic.
- One hour should be ample time to test yourself on the key ideas for a topic.
- Spread out the practice sessions for each topic – the optimum time to leave between each session is about one month but, if this isn't possible, just make the gaps as big as realistically possible.

Test Yourself

- Methods for testing yourself include: quizzes, practice questions, flashcards, past papers, explaining a topic to someone else, etc.
- This Complete Revision & Practice book provides seven practice opportunities per topic.
- Don't worry if you get an answer wrong – provided you check what the correct answer is, you are more likely to get the same or similar questions right in future!

Visit our website to download your free flashcards, for more information about the benefits of these techniques, and for further guidance on how to plan ahead and make them work for you.

www.collins.co.uk/collinsalevelrevision

Contents

Topic-Based Questions

Practice Exam Papers

Algebra and Functions

1 Given that $y = \frac{1}{32}x^5$, write $3y^{\frac{3}{5}}$ in the form ax^n. [3]

2 Show that $\frac{2 + 4\sqrt{8}}{\sqrt{8} - 4}$ can be written in the form $p + q\sqrt{2}$, where p and q are rational numbers, and hence write down the values of p and q. [4]

3 Factorise completely $2x^3 - 4x^2 - 30x$. [2]

4 Show that $(2x - 3y)(3x + 2y)(x - y)$ can be written in the form $ax^3 + bx^2y + cxy^2 + dy^3$ and hence write down the values of a, b, c and d. [3]

5 a) Show that $(x + 2)$ is a factor of $f(x) = 2x^3 + x^2 - 3x + 6$. [2]

   b) Hence, show that $x = -2$ is the only real root of the equation $f(x) = 0$. [4]

6 By considering the first four terms of the expansion of $(3 - \frac{x}{4})^8$ in ascending powers of x, find an estimate for 2.995^8. [7]

7 **a)** The function $f(x) = 9x^2 - 9kx + 3k$ has exactly one real root when $f(x) = 0$.

Find the possible values of k. [3]

b) Given $k > 0$, find the value of the root of $f(x) = 0$. [3]

8 Sketch the graph of $y = -x^2 - 2x - 2$. [5]

9 Show that the curve $f(x) = -x^2 + 3x - 6$ intersects the line $g(x) = -5x + 10$ at exactly one point and find the coordinates of that point. [5]

10 Solve the simultaneous equations $x^2 + 2xy + y^2 = 16$ and $2x + 3y = 8$. [5]

Algebra and Functions

11 Show the region on a coordinate grid and find the range of values of x for which

$y < x^2 + 3x - 4$ and $y > \frac{1}{2}x - 3$. [10]

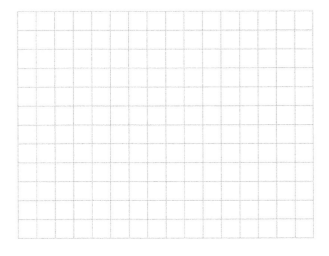

12 The graph shows $f(x) = 2x^2 + 10x - 23$ and $g(x) = -4x^2 - 2x + 25$.

Find the coordinates of the points of intersection and hence the range of values of x for which $g(x) > f(x)$. [5]

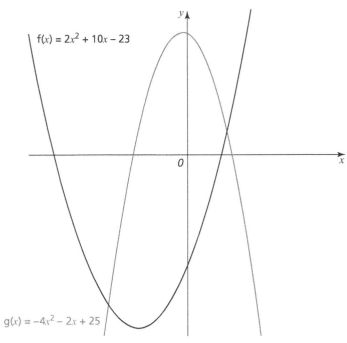

13 a) Use the Factor Theorem to show that $x = 3$ is a root of $f(x) = 0$ when
$f(x) = 2x^3 - 9x^2 + 7x + 6$. [1]

b) Sketch the graph of $y = f(x)$. [7]

14 $g(x) = (x + 2)(x - 3)(x - 1)(x - 4)$

a) Sketch the graph of $y = g(x)$. [3]

b) Write down the coordinates of the points of intersection of $g(x - 1)$ with the x- and the y-axis. [4]

15 Given the graph of $y = f(x)$ for $f(x) = \frac{3}{x^2}$,

express the equation $y = \frac{3}{(x - 1)^2} + 2$ in

the form $y = f(x + a) + b$ and hence

sketch the graph of $y = \frac{3}{(x - 1)^2} + 2$,

stating clearly the equations of the

asymptotes. [4]

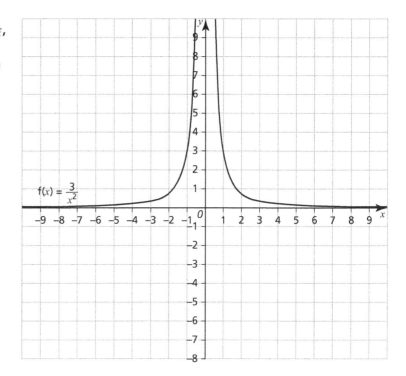

16 Given the curve $y = f(x)$ shown, sketch the
transformations $y = f(-x)$ and $y = -f(x)$. [4]

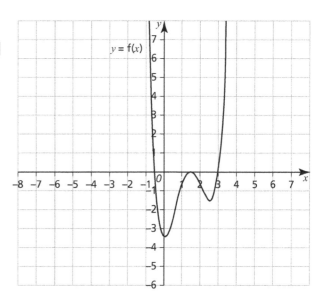

Total Marks _____ / 84

Coordinate Geometry

1 Find the equation of the line that passes through (–4, 8) and (–2, 5). [3]

2 **a)** Show that triangle ABC below is a right-angled triangle. [3]

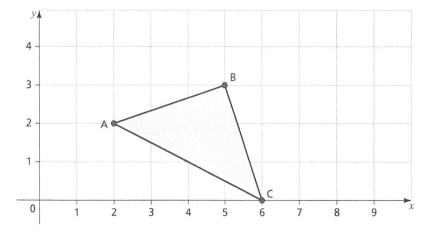

b) Find the area of triangle ABC. [3]

3 Line L_1 passes through points A(–1, 4) and B(2, 0). Find the equation of line L_2 that is parallel to L_1 and passes through point C(3, 2), giving your answer in the form $ax + by + c = 0$. [3]

4 Find the equation of the perpendicular bisector of line AB joining points A(1, 5) and B(4, 2). [5]

5 A(−3, 4), B(5, 8) and C(−2, 1), are points on a circle. Find the equation of the circle. [8]

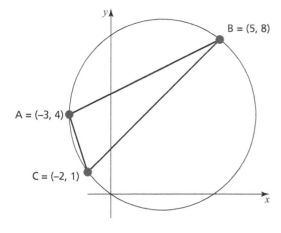

6 **a)** Given that (7, 0) lies on the circle with equation $(x - 4k)^2 + (y - k)^2 = 10$, find two possible values of k. [3]

b) Given that k is an integer, find the centre of the circle. [1]

7 The circle with equation $x^2 + y^2 - 4x - 6y = 2$ is shown.

a) Find the centre and the radius of the circle. [3]

b) Find the coordinates of points A and B, where the circle intersects the x-axis. [3]

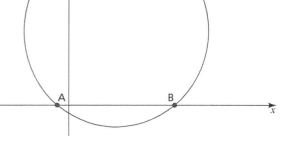

Total Marks _____ / 35

1

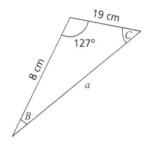

19 cm

127°

8 cm

a

B

C

a) Find the length of side a. [3]

b) Find the size of angle B. [3]

c) Find the size of angle C. [2]

2 Solve the equation $3\sin x + \cos x = 0$, $0° \leqslant x \leqslant 360°$. [4]

3 **a)** Show that $(1 + \sin x)^2 + \cos^2 x \equiv 2 + 2\sin x$. [3]

b) Hence solve the equation $(1 + \sin x)^2 + \cos^2 x = 1$, $0 \leqslant x \leqslant 2\pi$. [4]

4 **a)** Sketch the graphs of $y = 2\sin x$ and $y = 3\cos x$ on the same axes, $0° \leqslant x \leqslant 360°$. [3]

b) Write down the number of solutions there are for the equation $2\sin x = 3\cos x$, $0° \leqslant x \leqslant 360°$. [1]

Total Marks _____ / 23

Exponentials and Logarithms

1 Solve the equation $\log_2(3x - 1) = 4$. [3]

2 Solve the equation $\ln(5 - 2x) = 10$. [3]

3 Solve the equation $\log_{10}(x + 3) - \log_{10}(2x - 1) = 3$. [4]

4 Solve the equation $3^{1-x} = 2^x$. [4]

5 Sketch the graph of $y = -\ln(x + 1)$, showing any asymptotes and intersections with axes. [3]

6 Solve the equation $e^{2x} - 3e^x - 4 = 0$. [5]

7 Simplify the expression $4\log_4 2 + \log_4 8 - \log_4 2$. [3]

8 Solve the equation $2\log_2(x - 9) = \log_3 81 + 1$, giving your answers in surd form. [5]

9 Write $\ln\left(\dfrac{p^5}{q^6}\right)^{\frac{3}{2}}$ in the form $a\ln p + b\ln q$. [3]

10 Solve the equation $\log_6 z = 4\log_z 2 + 4\log_z 3$ [7]

Total Marks _____ / 40

Differentiation

1 **a)** Find the nature of the stationary points of the curve with equation $y = \frac{1}{4}x^4 + \frac{1}{3}x^3 - 6x^2 + 3$. **[5]**

b) Hence sketch the curve with equation $y = \frac{1}{4}x^4 + \frac{1}{3}x^3 - 6x^2 + 3$. **[3]**

2 Differentiate with respect to x: $3x^3 - \sqrt{x} + \dfrac{x^3 + 4x}{x^3}$ **[4]**

3 A closed cylinder has volume 200 cm³.

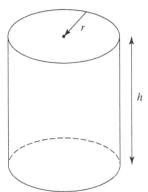

a) Show that the surface area = $2\pi r^2 + \frac{400}{r}$ [4]

b) Use calculus to find the maximum value of the surface area. [4]

4 Find the equation of the tangent to the curve with equation $y = x^2 - 7x + 10$ at the point (2, 0). [4]

Total Marks _____ / 24

1 The diagram below is a sketch of the curve with equation $y = x - x^2$.

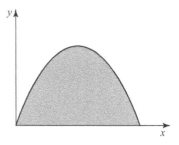

Find the area of the shaded region. [5]

2 **a)** Find $\int (x^2 - x)\left(x^{\frac{1}{2}} + 2\right)dx$. [4]

b) Use your answer to part **a)** to evaluate $\int_1^3 (x^2 - x)\left(x^{\frac{1}{2}} + 2\right)dx$. [2]

3 Given that $y^{\frac{1}{3}} = 3x + 5$

a) Find y. [2]

b) Find $\int y\,dx$. [3]

4 The gradient of a curve is given by $f'(x) = 3x^2 + 4x - 3x^{-\frac{1}{2}}$

Given that the curve passes through the point (1, 4), find the equation of the curve in the form $y = f(x)$. [5]

Total Marks _____ / 21

Vectors

1 A is the point (0, 10) and B is the point (12, 26). Find the coordinates of the point C, which splits the line AB in the ratio 1 : 4. [6]

2 Find an expression for the distance between the points ($-3a$, 2) and (4, $-5a$), where a is a constant. Hence show that the distance is a maximum when $a = -\frac{11}{17}$ [6]

3 A parallelogram ABCD has coordinates A(-2, -8), B(1, 5) and C(0, 7).

Find the coordinates of D. [4]

4 A vector **a** has magnitude 24 and acts in a direction of 120° to the horizontal. Express **a** in Cartesian vector form. [2]

5 Find a vector of magnitude $2\sqrt{5}$ units in the direction of ($-\mathbf{i} + 3\mathbf{j}$). [4]

6 In a triangle OAB, D splits OA in the ratio 1 : 3 and C splits AB in the ratio 1 : 3. E is the point of intersection of lines OC and BD. Given $\overrightarrow{OA}$ = **a** and $\overrightarrow{OB}$ = **b**, find $\overrightarrow{OE}$ in terms of **a** and **b**.

[11]

7 **a** and **b** are non-zero and non-parallel vectors. Given vectors **p** = 3**a** + (2λ + 1)**b** and **q** = (2 − 5λ)**a** − 12**b** are parallel, where λ is an integer, determine the value of λ.

[5]

8 Find the magnitude and direction of the vector **a** = 2**i** − **j**.

[4]

9 The vector **a** has magnitude 8 and **b** has magnitude 14. The angle between the two vectors is 30°. Find |**R**|, the magnitude of the resultant of **a** and **b**, and the angle **R** makes with **b**.

[6]

Total Marks _____ / 48

Proof

1 The equation $y = x^2 + 2kx + 1 = 0$ has two equal roots. Show that $k = 1$ or -1. [3]

2 Prove that $2x^3 + x^2 - 18x - 9 = (x - 3)(2x + 1)(x + 3)$. [5]

3 Use completing the square to prove that $y^2 - 5y + 30$ is positive for all values of y. [3]

Total Marks .... / 11

Statistical Sampling & Statistical Hypothesis Testing

1 A single observation X is taken from a binomial distribution B(40, p). The observation is used to test the hypothesis H$_0$: $p = 0.3$ against H$_1$: $p \neq 0.3$.

    a) Using a 5% significance level, find the critical region for this test. [3]

    b) Write down the actual significance level of this test. [1]

    c) A value of X obtained is 5. State a conclusion that can be drawn based on this value, giving a reason for your answer. [2]

2 A random variable $X \sim$ B(30, 0.5).

    a) Find P($X \leqslant 8$). [1]

    b) Find P($X \geqslant 21$). [2]

Total Marks _____ / 9

1 Meghan wants to investigate the average rainfall in Leuchars. The extract below shows her sample taken from the large data set.

Date	LEUCHARS Daily Mean Temperature (0900–0900) (°C)	© Crown Copyright Met Office 2015 Daily Total Rainfall (0900–0900) (mm)	Daily Total Sunshine (0000–2400) (hrs)	Daily Mean Wind Speed (0000–2400) (kn)
01/05/2015	3.8	tr	9.1	8
02/05/2015	4.6	7.2	5.7	13
03/05/2015	9.3	15.4	0	15
04/05/2015	11.0	3.8	8.6	10
05/05/2015	7.5	16	1.3	12
06/05/2015	9.4	1	2.1	7
07/05/2015	6.8	tr	7.2	11
08/05/2015	4.1	5.6	5	7
09/05/2015	9.2	2	5.2	8
10/05/2015	9.8	3	0.4	11
11/05/2015	12.7	0.6	9.9	18
12/05/2015	11.5	tr	8.5	20
13/05/2015	9.5	tr	9.6	10
14/05/2015	9.1	0	0	8
15/05/2015	7.4	0.6	6.4	11
16/05/2015	9.6	0.2	12.2	16
17/05/2015	9.9	0.8	9.9	16
18/05/2015	10.0	2.4	2.6	8
19/05/2015	8.7	0.6	7.4	9
20/05/2015	11.9	1	9.1	10

a) Find the median daily mean temperature. [2]

b) An outlier is an observation which falls either 1.5 × interquartile range above the upper quartile or 1.5 × interquartile range below the lower quartile. Determine if any of the mean daily temperature values are outliers. [3]

c) Draw a box plot to show the mean daily temperature. [3]

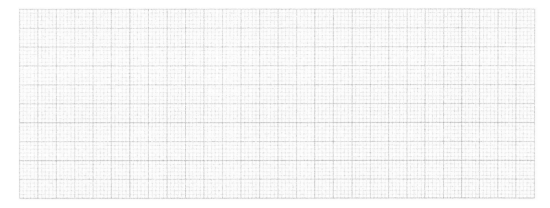

Daniel believes there is a linear relationship between the mean daily temperature and the total daily sunshine.

d) Draw a scatter diagram and comment on Daniel's claim. [4]

Data Presentation and Interpretation

2 The cumulative frequency diagram shows the distribution of speeds of 60 motorists on motorway A and motorway B.

Compare the speeds of motorway A and motorway B. [4]

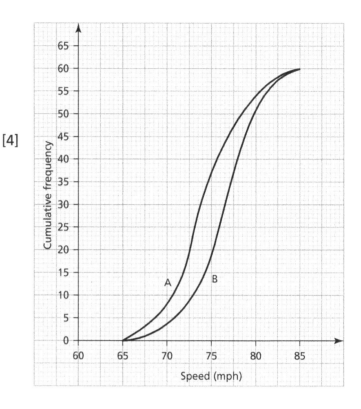

3 The table shows the lengths of leaves from a tree measured to the nearest millimetre.

Length (mm)	Frequency
125–139	15
140–149	25
150–154	19
155–169	8

a) Calculate an estimate for the mean length. [3]

b) Calculate an estimate for the standard deviation. [3]

c) State a reason to justify representing this data in a histogram. [1]

d) Draw a histogram to represent this data. [4]

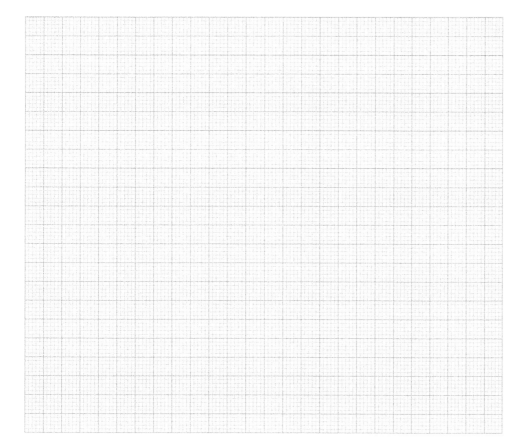

Total Marks .... / 27

Probability

1 A bag contains 16 coloured counters: 8 coloured purple, 5 coloured orange and 3 coloured red.

Two counters are drawn from the bag without replacement.

a) Draw a tree diagram to represent this information. [4]

b) Find the probability that both counters are orange. [2]

c) Find the probability that the two counters are the same colour. [4]

2 The Venn diagram shows the probabilities that a group of children like carrots (A) or peas (B).

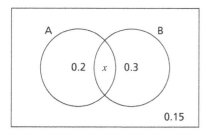

a) Find the value of x. [3]

b) A child is selected at random.

Find the probability that this child likes carrots given that they like peas. [3]

c) Determine whether or not the events "like peas" and "like carrots" are independent. [4]

Total Marks _____ / 20

The Language of Mechanics

1 A force of 150 N acting upon a mass, m, produces an acceleration of 2.5 ms^{-2}.

Find the value of m. [2]

2 A particle lies on the x-axis at the point where $x = 15$. After 12 seconds, it is at the point where $x = -3$. Find the velocity of the particle, assumed to be constant. [3]

3 A particle of mass 20 kg is pulled along a straight line, with a constant force of 26 N. [3]

If its acceleration is 0.6 ms^{-2}, find the resistive force on the particle, assumed constant.

4 A force of 20 N acts on a stationary 8 kg mass. Find its velocity after 10 seconds. [4]

5 An object of mass 12 kg has its velocity increased from 0.5 ms^{-1} to 2.5 ms^{-1} in one minute.

Find the force acting on the object, if assumed to be constant. [4]

Total Marks _____ / 16

Kinematics

1 The velocity–time diagram illustrates the journey of a cyclist. Using the diagram, find:

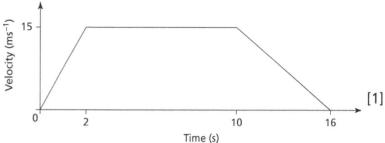

a) the initial acceleration of the cyclist [1]

b) the final deceleration of the cyclist [1]

c) the total distance travelled [2]

d) the cyclist's average speed. [2]

2 A car has initial velocity of $40\,ms^{-1}$, undergoing constant deceleration to reach speed V after 20 seconds.

a) Draw a velocity–time diagram to illustrate the journey of the car. [4]

b) Given that the car has travelled a distance of 550 m, find V. [2]

c) Find the car's speed after 10 seconds. [2]

Kinematics

3 A train sets off from rest, accelerating at $0.4\,ms^{-2}$ until reaching a speed of $16\,ms^{-1}$. It maintains this speed for T seconds before decelerating at a rate of $\frac{4}{15}\,ms^{-2}$, until it comes to a halt.

a) Draw a velocity–time graph showing the journey of the train. [5]

b) Given that the train travels 1920 m, determine the value of T. [3]

4 A particle is projected in a horizontal direction with initial speed $10\,ms^{-1}$. If its acceleration is $-0.25\,ms^{-2}$, find the time(s) when its displacement is 30 m from its starting position. [5]

5 A car is travelling at $15\,ms^{-1}$ when the driver sees traffic lights ahead. She brakes, and decelerates at a constant rate of $3.5\,ms^{-2}$, coming to rest at the lights. Find:

a) the distance travelled by the car [2]

b) the time for which the driver was braking. [2]

6 An object travels in a straight line such that after t seconds, its displacement s from its starting point is given by $s = \dfrac{t^4}{2} - \dfrac{17t^3}{3} + \dfrac{35t^2}{2}$

Find s, when the object first comes to an instantaneous rest. [7]

7 A particle moves in a straight line with acceleration $a = 2t^2 - 1$.

Given its velocity is 6 ms^{-1} when $t = 3$, find its velocity when $t = 4$. [5]

8 A particle moves in a straight line such that its velocity at time t is given by $v = 5 - 5t$.

Find the total **distance travelled** during the first two seconds of its motion. [5]

9 A particle moves in a straight line such that at time t its acceleration is given by $a = 2t$.
At $t = 3$ its displacement is 10 m and at $t = 6$ its displacement is 70 m.

Find an expression for its displacement, s, in terms of t. [7]

Total Marks _____ / 55

Forces and Newton's Laws

1 A particle is projected vertically upwards with speed 40 ms⁻¹.

Find the maximum height reached by the particle. [3]

2 A particle has mass 4 kg. Find the force required to change the particle's velocity from $(-3\mathbf{i} - 9\mathbf{j})$ ms⁻¹ to $(4\mathbf{i} + 8\mathbf{j})$ ms⁻¹ in two seconds. [4]

3 A car of mass 1200 kg pulls a caravan of mass 800 kg along a straight horizontal road. Both car and caravan encounter resistive forces proportional to their masses. The car exerts a driving force of 1.1 kN, producing an acceleration of 0.5 ms⁻².

Find the tension in the rope (assumed light). [8]

4 A particle is dropped from the top of a cliff, 100 m high. At the same time, a particle is projected vertically upwards from the foot of the cliff with a speed of 40 ms⁻¹.

Find the distance from the foot of the cliff that the particles meet. [8]

5 Forces $\mathbf{p} = \begin{pmatrix} a \\ 2b \end{pmatrix}$, $\mathbf{q} = \begin{pmatrix} b \\ 10 \end{pmatrix}$ and $\mathbf{r} = \begin{pmatrix} 3 \\ a \end{pmatrix}$ act on a particle that remains in equilibrium.

Find the value of a and the value of b. [6]

6 A particle of mass 3 kg is joined by a light string to another particle of mass 2 kg. The 2 kg mass is joined by another light string to a 1 kg mass. The 3 kg mass is pulled along a smooth horizontal floor with force 18 N.

Find the acceleration of the system and the tension in each of the strings. [8]

7 Two particles, A and B, of masses 5m kg and 6m kg respectively, are joined by a light string, hanging over a smooth pulley. Initially they are at rest at the same horizontal level, with the strings taut.

a) Find the acceleration of the system. [6]

b) Find the tension in the string. [2]

c) Find the force on the pulley. [2]

d) Find the distance travelled by B in the first two seconds. [2]

8 Two particles, A and B, of masses $5m$ kg and $6m$ kg respectively, are joined by a light string. Particle A rests on a smooth horizontal table, with particle B hanging vertically from one end of the table, passing over a smooth fixed pulley. Particle A is a distance of 1.5 m from the pulley.

The system is initially at rest, then released.

a) Find the acceleration of the system. [6]

b) Find the tension in the string. [2]

c) Find the force on the pulley. [2]

d) Find the time taken for particle A to bang into the pulley. [2]

Total Marks _____ / 61

Collins

GCE
Mathematics
Advanced Subsidiary
Paper 1: Pure Mathematics

Time: 2 hours

You must have:
Mathematical Formulae and Statistical Tables
Calculator

Calculators must not have the facility for algebraic manipulation, differentiation and integration, or have retrievable mathematical formulae stored in them.

Instructions
- Use **black** ink or ball-point pen.
- If pencil is used for diagrams/sketches/graphs, it must be dark (HB or B).
- Answer **all** the questions and ensure that your answers to parts of questions are clearly labelled.
- Write your answers in the spaces provided.
- Carry out your working on separate sheets of paper where necessary. You should show sufficient working to make your methods clear. Answers without working may not gain full credit.
- Inexact answers should be given to three significant figures unless otherwise stated.

Information
- Mathematical Formulae and Statistical Tables are provided on pages 219–224.
- There are 16 questions in this question paper. The total mark for this paper is 100.
- The marks for each question are shown in brackets
  – *use this as a guide as to how much time to spend on each question.*

Advice
- Read each question carefully before you start to answer it.
- Try to answer every question.
- Check your answers if you have time at the end.
- If you change your mind about an answer, cross it out and put your new answer underneath.

Name: ....

Answer ALL questions. Write your answers in the spaces provided.

1. Solve the simultaneous equations $y = 2x^2 - 3x + 5$ and $y = 8 - 2x$ (4)

(Total for Question 1 is 4 marks)

2. Simplify the expression $\dfrac{\sqrt{3} - 1}{\sqrt{3} + 1} + \dfrac{6}{\sqrt{3}}$, giving your answer in the form $a + b\sqrt{3}$, where a and b are integers. (4)

(Total for Question 2 is 4 marks)

3. The curve $y = f(x)$ passes through the point $(-2, -5)$.

Given $f'(x) = x^3 - 3x^2 + x + 6$, find the equation of the curve $y = f(x)$. (5)

(Total for Question 3 is 5 marks)

4. Consider the curve $y = 3 - \dfrac{4}{\sqrt{x}}$

(a) Show that the point $(4, 1)$ lies on the curve. (2)

(b) Find the equation of the normal to the curve at this point, expressing your answer in the form $ax + by + c = 0$ (6)

(Total for Question 4 is 8 marks)

5. Find the first three terms, in ascending powers of x, of the binomial expansion of $(3 - 2x)^9$, giving each term in its simplest form. (6)

(Total for Question 5 is 6 marks)

6. The quadratic equation $3x^2 - (k + 1)x + 1 = 0$ has no real roots.

Show that $a + b\sqrt{3} < k < c + d\sqrt{3}$, where a, b, c and d are constants to be determined. (5)

(Total for Question 6 is 5 marks)

7. Solve the equation $\log_6 x + \log_6(x - 9) = 2$ (6)

(Total for Question 7 is 6 marks)

8. Calculate (to 3 significant figures) the area of the triangle ABC, where $AB = 13\,\text{cm}$, $AC = 25\,\text{cm}$ and $\angle ABC = 135°$ (5)

(Total for Question 8 is 5 marks)

9. It is conjectured that $\tan(A + B) \equiv \tan A + \tan B$, when $\tan A$, $\tan B$ and $\tan(A + B)$ each take finite values.

   (a) Prove by counter example that this is not true. (2)

   (b) State possible A and B value(s) for which the conjecture holds. (2)

   Note: Any answers involving the use of a calculator in this question score no marks.

(Total for Question 9 is 4 marks)

10. Solve the equation $10\cos^2 x + \sin x - 8 = 0$ in the range $-180° < x < 360°$. (8)

(Total for Question 10 is 8 marks)

11. Given point A has position vector $3\mathbf{i} + 12\mathbf{j}$ and point B has position vector $7\mathbf{i} + 10\mathbf{j}$, find a vector of magnitude 10 in the direction of AB, giving your answer in surd form.　(5)

(Total for Question 11 is 5 marks)

12. Prove, from first principles, that the derivative of $2x^3$ (with respect to x) is $6x^2$.　(5)

(Total for Question 12 is 5 marks)

13. Show that the function $y = \dfrac{19 - 5x}{x - 3}$ may be written in the form $y = \dfrac{4}{x + a} + b$.

Hence or otherwise, sketch the graph of $y = \dfrac{19 - 5x}{x - 3}$, marking any asymptotes and intersections with axes clearly.

(7)

(Total for Question 13 is 7 marks)

14. Consider the cubic expression $f(x) = -x^3 - x^2 + 10x - 8$.

(a) (i) Show that $(x + 4)$ is a factor of $f(x)$. (2)

(ii) Hence or otherwise, factorise $f(x)$ completely. (4)

(b) Sketch the curve $y = f(x)$. (1)

(c) Shade an area on your graph illustrating the region satisfying $y \geqslant -x^3 - x^2 + 10x - 8$, $x \leqslant 0$ **and** $4y < -20 - 5x$. (3)

(Total for Question 14 is 10 marks)

15. The equation of a circle is given by $4x^2 + 4y^2 - 12x - 24y + 41 = 0$.

(a) Find the coordinates of the centre of the circle and its radius. (5)

(b) Two separate tangents to the circle each pass through the origin.
Find the equations of these tangents. (5)

(Total for Question 15 is 10 marks)

16. The diagram below shows the graph of $y = 1 + e^{-2x}$

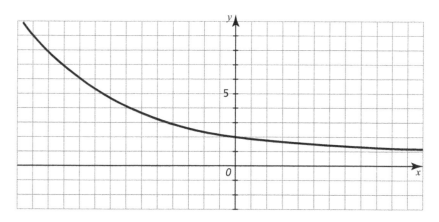

(a) This graph may be produced from the graph of $y = e^x$ through a series of three transformations. State each of these in turn. (3)

(b) Show that the area bounded by the x-axis, the curve $y = 1 + e^{-2x}$, and the lines $x = -\ln 3$ and $x = \ln 2$ may be written in the form $a + b\ln 6$, where a and b are to be determined. (5)

Note: You may use the result $\int e^{kx} dx = \frac{1}{k}e^{kx}(+c)$

(Total for Question 16 is 8 marks)

TOTAL FOR PAPER IS 100 MARKS

THIS PAGE HAS INTENTIONALLY
BEEN LEFT BLANK

GCE
Mathematics
Advanced Subsidiary
Paper 2: Statistics and Mechanics

Time: 1 hour 15 minutes

You must have:
Mathematical Formulae and Statistical Tables
Calculator

Calculators must not have the facility for algebraic manipulation, differentiation and integration, or have retrievable mathematical formulae stored in them.

Instructions
- Use **black** ink or ball-point pen.
- If pencil is used for diagrams/sketches/graphs, it must be dark (HB or B).
- There are **two** sections in this question paper. Answer **all** the questions in Section A and **all** the questions in Section B.
- Write your answers in the spaces provided.
- Carry out your working on separate sheets of paper where necessary. You should show sufficient working to make your methods clear. Answers without working may not gain full credit.
- Inexact answers should be given to three significant figures unless otherwise stated.

Information
- Mathematical Formulae and Statistical Tables are provided on pages 219–224.
- There are 9 questions in this question paper. The total mark for this paper is 60.
- The marks for each question are shown in brackets
  – *use this as a guide as to how much time to spend on each question.*

Advice
- Read each question carefully before you start to answer it.
- Try to answer every question.
- Check your answers if you have time at the end.
- If you change your mind about an answer, cross it out and put your new answer underneath.

Name: ....

SECTION A: STATISTICS

1. The mean value of six positive numbers, $a, b, 2, 2, 3$ and 5, is $\frac{8}{3}$ and their variance is $\frac{14}{9}$.

   Given $a < b$, find the value of a and the value of b. (6)

(Total for Question 1 is 6 marks)

2. The Venn diagram below illustrates the probabilities that students in Wilmington High School take either mathematics (M), economics (E) or physics (P). p and q represent probabilities.

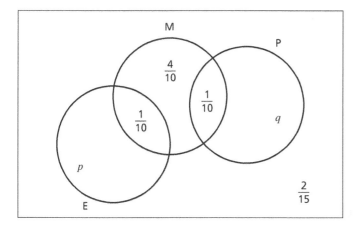

(a) Explain why events P and E cannot be independent. (1)

(b) Given that the probability that a student takes mathematics is independent from the probability that a student takes economics, find the value of p. (2)

(c) Find the value of q. (2)

(Total for Question 2 is 5 marks)

3. For the data set below, identify any outliers and draw a box plot.

    (An outlier is defined as a value more than 1.5 times the interquartile range below Q_1 or above Q_3).

    16, 17, 12, 31, 9, 19, 18, 18, 13, 27, 15   (5)

(Total for Question 3 is 5 marks)

4. The following is a cumulative frequency diagram for the daily mean temperature occurring at Leeming during the month of June 1987. The raw data was originally taken from the large data set.

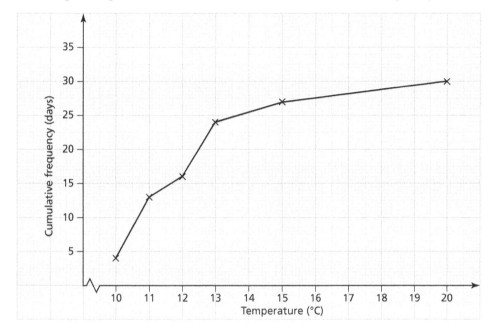

(a) Estimate the number of days in June 1987 when the mean temperature lay between 12.5°C and 17.5°C. (2)

(b) Estimate the probability that a randomly chosen day in June 1987 had a mean temperature of less than 10.5°C. (2)

(c) Explain why, from the information given, it would be difficult to provide a reliable estimate for the range of mean temperatures during the month of June 1987. (1)

(Total for Question 4 is 5 marks)

5. 12% of year 11 students at Wilmington High School express an intention to leave school at the end of the year and attend the local sixth-form college.

In a sample of 20 students chosen at random from the year, find (to 3 significant figures) the probability that the number of students who intend to leave is:

(a) (i) precisely 3 (2)

(ii) 3 or more. (2)

The following year, it was noted that 5 students from a random sample of 20 students from year 11 intended to leave for the local college.

(b) Determine, at the 5% significance level, whether there has been an increase in the number of year 11 students intending to leave the school. State your hypotheses clearly. (5)

(Total for Question 5 is 9 marks)

TOTAL FOR SECTION A IS 30 MARKS

SECTION B: MECHANICS

6. A cyclist accelerates from rest for 6 seconds before reaching her maximum velocity of 12 ms^{-1}. She maintains this speed for another 30 seconds, before decelerating at a rate of 0.8 ms^{-2}.

   (a) Draw a velocity–time graph to illustrate the cyclist's motion. (1)

   (b) Find the total time for the cyclist's journey. (2)

   (c) Find the total distance travelled by the cyclist. (2)

(Total for Question 6 is 5 marks)

7. An aeroplane accelerates from rest at $2.5\,\mathrm{ms^{-2}}$ along a horizontal runway, reaching a speed at take-off of $70\,\mathrm{ms^{-1}}$. It uses the full length of the runway.

(a) Find the length of the runway. (2)

(b) Find the time it takes the aeroplane to travel half the distance of the runway. (2)

(Total for Question 7 is 4 marks)

8. A particle travels in a straight line with acceleration $a = 4 - 2t, 0 \leqslant t < 5$.
   Its velocity is initially $\frac{3}{2}$ ms⁻¹.

   (a) Find the particle's velocity, v, in terms of t. (2)

   (b) Find the speed of the particle when $t = 5$. (2)

   (c) Find the time when the particle is instantaneously at rest. (2)

   (d) Find the total distance travelled by the particle. (3)

(Total for Question 8 is 9 marks)

9. Two particles, A and B, are attached to the ends of a light inextensible string and have masses $5m$ and $2m$ respectively. The string passes over a smooth pulley, fixed above a horizontal floor. The particles are held at rest at a height of $3\,\text{m}$ above the floor. The particles are then released from rest.

(a) Write down equations of motion for both particle A and particle B. (3)

(b) Find the acceleration of the masses immediately after the particles are released. (2)

(c) Find the tension in the string. (1)

(d) Find the maximum height above the ground reached by particle B. (5)

(e) State how you have used the fact that the string is light in the question. (1)

(Total for Question 9 is 12 marks)

TOTAL FOR SECTION B IS 30 MARKS

TOTAL FOR PAPER IS 60 MARKS

Mathematical Formulae

Pure Mathematics

Mensuration

Surface area of sphere $= 4\pi r^2$

Area of curved surface of cone $= \pi r \times$ slant height

Binomial series

$$(a+b)^n = a^n + \binom{n}{1}a^{n-1}b + \binom{n}{2}a^{n-2}b^2 + \ldots + \binom{n}{r}a^{n-r}b^r + \ldots + b^n \ (n \in \mathbb{N})$$

where $\binom{n}{r} = {}^nC_r = \dfrac{n!}{r!(n-r)!}$

Logarithms and exponentials

$$\log_a x = \frac{\log_b x}{\log_b a}$$

$$e^{x\ln a} = a^x$$

Differentiation

First Principles

$$f'(x) = \lim_{h \to 0} \frac{f(x+h) - f(x)}{h}$$

Statistics

Probability

$P(A') = 1 - P(A)$

Standard deviation

Standard deviation $= \sqrt{(\text{Variance})}$

Interquartile range $= \text{IQR} = Q_3 - Q_1$

For a set of n values $x_1, x_2, \ldots x_i, \ldots x_n$

$$S_{xx} = \Sigma\left(x_i - \bar{x}\right)^2 = \Sigma x_i^2 - \frac{\left(\Sigma x_i\right)^2}{n}$$

$$\text{Standard deviation} = \sqrt{\frac{S_{xx}}{n}} \text{ or } \sqrt{\frac{\Sigma x^2}{n} - \bar{x}^2}$$

Mechanics

Kinematics

For motion in a straight line with constant acceleration:

$v = u + at$

$s = ut + \frac{1}{2}at^2$

$s = vt - \frac{1}{2}at^2$

$v^2 = u^2 + 2as$

$s = \frac{1}{2}(u + v)t$

Statistical Tables

Binomial Cumulative Distribution Function

The tabulated value is $P(X \leqslant x)$, where X has a binomial distribution with index n and parameter p.

$p =$	0.05	0.10	0.15	0.20	0.25	0.30	0.35	0.40	0.45	0.50
$n = 5, x = 0$	0.7738	0.5905	0.4437	0.3277	0.2373	0.1681	0.1160	0.0778	0.0503	0.0312
1	0.9774	0.9185	0.8352	0.7373	0.6328	0.5282	0.4284	0.3370	0.2562	0.1875
2	0.9988	0.9914	0.9734	0.9421	0.8965	0.8369	0.7648	0.6826	0.5931	0.5000
3	1.0000	0.9995	0.9978	0.9933	0.9844	0.9692	0.9460	0.9130	0.8688	0.8125
4	1.0000	1.0000	0.9999	0.9997	0.9990	0.9976	0.9947	0.9898	0.9815	0.9688
$n = 6, x = 0$	0.7351	0.5314	0.3771	0.2621	0.1780	0.1176	0.0754	0.0467	0.0277	0.0156
1	0.9672	0.8857	0.7765	0.6554	0.5339	0.4202	0.3191	0.2333	0.1636	0.1094
2	0.9978	0.9842	0.9527	0.9011	0.8306	0.7443	0.6471	0.5443	0.4415	0.3438
3	0.9999	0.9987	0.9941	0.9830	0.9624	0.9295	0.8826	0.8208	0.7447	0.6563
4	1.0000	0.9999	0.9996	0.9984	0.9954	0.9891	0.9777	0.9590	0.9308	0.8906
5	1.0000	1.0000	1.0000	0.9999	0.9998	0.9993	0.9982	0.9959	0.9917	0.9844
$n = 7, x = 0$	0.6983	0.4783	0.3206	0.2097	0.1335	0.0824	0.0490	0.0280	0.0152	0.0078
1	0.9556	0.8503	0.7166	0.5767	0.4449	0.3294	0.2338	0.1586	0.1024	0.0625
2	0.9962	0.9743	0.9262	0.8520	0.7564	0.6471	0.5323	0.4199	0.3164	0.2266
3	0.9998	0.9973	0.9879	0.9667	0.9294	0.8740	0.8002	0.7102	0.6083	0.5000
4	1.0000	0.9998	0.9988	0.9953	0.9871	0.9712	0.9444	0.9037	0.8471	0.7734
5	1.0000	1.0000	0.9999	0.9996	0.9987	0.9962	0.9910	0.9812	0.9643	0.9375
6	1.0000	1.0000	1.0000	1.0000	0.9999	0.9998	0.9994	0.9984	0.9963	0.9922
$n = 8, x = 0$	0.6634	0.4305	0.2725	0.1678	0.1001	0.0576	0.0319	0.0168	0.0084	0.0039
1	0.9428	0.8131	0.6572	0.5033	0.3671	0.2553	0.1691	0.1064	0.0632	0.0352
2	0.9942	0.9619	0.8948	0.7969	0.6785	0.5518	0.4278	0.3154	0.2201	0.1445
3	0.9996	0.9950	0.9786	0.9437	0.8862	0.8059	0.7064	0.5941	0.4770	0.3633
4	1.0000	0.9996	0.9971	0.9896	0.9727	0.9420	0.8939	0.8263	0.7396	0.6367
5	1.0000	1.0000	0.9998	0.9988	0.9958	0.9887	0.9747	0.9502	0.9115	0.8555
6	1.0000	1.0000	1.0000	0.9999	0.9996	0.9987	0.9964	0.9915	0.9819	0.9648
7	1.0000	1.0000	1.0000	1.0000	1.0000	0.9999	0.9998	0.9993	0.9983	0.9961
$n = 9, x = 0$	0.6302	0.3874	0.2316	0.1342	0.0751	0.0404	0.0207	0.0101	0.0046	0.0020
1	0.9288	0.7748	0.5995	0.4362	0.3003	0.1960	0.1211	0.0705	0.0385	0.0195
2	0.9916	0.9470	0.8591	0.7382	0.6007	0.4628	0.3373	0.2318	0.1495	0.0898
3	0.9994	0.9917	0.9661	0.9144	0.8343	0.7297	0.6089	0.4826	0.3614	0.2539
4	1.0000	0.9991	0.9944	0.9804	0.9511	0.9012	0.8283	0.7334	0.6214	0.5000
5	1.0000	0.9999	0.9994	0.9969	0.9900	0.9747	0.9464	0.9006	0.8342	0.7461
6	1.0000	1.0000	1.0000	0.9997	0.9987	0.9957	0.9888	0.9750	0.9502	0.9102
7	1.0000	1.0000	1.0000	1.0000	0.9999	0.9996	0.9986	0.9962	0.9909	0.9805
8	1.0000	1.0000	1.0000	1.0000	1.0000	1.0000	0.9999	0.9997	0.9992	0.9980
$n = 10, x = 0$	0.5987	0.3487	0.1969	0.1074	0.0563	0.0282	0.0135	0.0060	0.0025	0.0010
1	0.9139	0.7361	0.5443	0.3758	0.2440	0.1493	0.0860	0.0464	0.0233	0.0107
2	0.9885	0.9298	0.8202	0.6778	0.5256	0.3828	0.2616	0.1673	0.0996	0.0547
3	0.9990	0.9872	0.9500	0.8791	0.7759	0.6496	0.5138	0.3823	0.2660	0.1719
4	0.9999	0.9984	0.9901	0.9672	0.9219	0.8497	0.7515	0.6331	0.5044	0.3770

$p =$	0.05	0.10	0.15	0.20	0.25	0.30	0.35	0.40	0.45	0.50
5	1.0000	0.9999	0.9986	0.9936	0.9803	0.9527	0.9051	0.8338	0.7384	0.6230
6	1.0000	1.0000	0.9999	0.9991	0.9965	0.9894	0.9740	0.9452	0.8980	0.8281
7	1.0000	1.0000	1.0000	0.9999	0.9996	0.9984	0.9952	0.9877	0.9726	0.9453
8	1.0000	1.0000	1.0000	1.0000	1.0000	0.9999	0.9995	0.9983	0.9955	0.9893
9	1.0000	1.0000	1.0000	1.0000	1.0000	1.0000	1.0000	0.9999	0.9997	0.9990
$n = 12, x = 0$	0.5404	0.2824	0.1422	0.0687	0.0317	0.0138	0.0057	0.0022	0.0008	0.0002
1	0.8816	0.6590	0.4435	0.2749	0.1584	0.0850	0.0424	0.0196	0.0083	0.0032
2	0.9804	0.8891	0.7358	0.5583	0.3907	0.2528	0.1513	0.0834	0.0421	0.0193
3	0.9978	0.9744	0.9078	0.7946	0.6488	0.4925	0.3467	0.2253	0.1345	0.0730
4	0.9998	0.9957	0.9761	0.9274	0.8424	0.7237	0.5833	0.4382	0.3044	0.1938
5	1.0000	0.9995	0.9954	0.9806	0.9456	0.8822	0.7873	0.6652	0.5269	0.3872
6	1.0000	0.9999	0.9993	0.9961	0.9857	0.9614	0.9154	0.8418	0.7393	0.6128
7	1.0000	1.0000	0.9999	0.9994	0.9972	0.9905	0.9745	0.9427	0.8883	0.8062
8	1.0000	1.0000	1.0000	0.9999	0.9996	0.9983	0.9944	0.9847	0.9644	0.9270
9	1.0000	1.0000	1.0000	1.0000	1.0000	0.9998	0.9992	0.9972	0.9921	0.9807
10	1.0000	1.0000	1.0000	1.0000	1.0000	1.0000	0.9999	0.9997	0.9989	0.9968
11	1.0000	1.0000	1.0000	1.0000	1.0000	1.0000	1.0000	1.0000	0.9999	0.9998
$n = 15, x = 0$	0.4633	0.2059	0.0874	0.0352	0.0134	0.0047	0.0016	0.0005	0.0001	0.0000
1	0.8290	0.5490	0.3186	0.1671	0.0802	0.0353	0.0142	0.0052	0.0017	0.0005
2	0.9638	0.8159	0.6042	0.3980	0.2361	0.1268	0.0617	0.0271	0.0107	0.0037
3	0.9945	0.9444	0.8227	0.6482	0.4613	0.2969	0.1727	0.0905	0.0424	0.0176
4	0.9994	0.9873	0.9383	0.8358	0.6865	0.5155	0.3519	0.2173	0.1204	0.0592
5	0.9999	0.9978	0.9832	0.9389	0.8516	0.7216	0.5643	0.4032	0.2608	0.1509
6	1.0000	0.9997	0.9964	0.9819	0.9434	0.8689	0.7548	0.6098	0.4522	0.3036
7	1.0000	1.0000	0.9994	0.9958	0.9827	0.9500	0.8868	0.7869	0.6535	0.5000
8	1.0000	1.0000	0.9999	0.9992	0.9958	0.9848	0.9578	0.9050	0.8182	0.6964
9	1.0000	1.0000	1.0000	0.9999	0.9992	0.9963	0.9876	0.9662	0.9231	0.8491
10	1.0000	1.0000	1.0000	1.0000	0.9999	0.9993	0.9972	0.9907	0.9745	0.9408
11	1.0000	1.0000	1.0000	1.0000	1.0000	0.9999	0.9995	0.9981	0.9937	0.9824
12	1.0000	1.0000	1.0000	1.0000	1.0000	1.0000	0.9999	0.9997	0.9989	0.9963
13	1.0000	1.0000	1.0000	1.0000	1.0000	1.0000	1.0000	1.0000	0.9999	0.9995
14	1.0000	1.0000	1.0000	1.0000	1.0000	1.0000	1.0000	1.0000	1.0000	1.0000
$n = 20, x = 0$	0.3585	0.1216	0.0388	0.0115	0.0032	0.0008	0.0002	0.0000	0.0000	0.0000
1	0.7358	0.3917	0.1756	0.0692	0.0243	0.0076	0.0021	0.0005	0.0001	0.0000
2	0.9245	0.6769	0.4049	0.2061	0.0913	0.0355	0.0121	0.0036	0.0009	0.0002
3	0.9841	0.8670	0.6477	0.4114	0.2252	0.1071	0.0444	0.0160	0.0049	0.0013
4	0.9974	0.9568	0.8298	0.6296	0.4148	0.2375	0.1182	0.0510	0.0189	0.0059
5	0.9997	0.9887	0.9327	0.8042	0.6172	0.4164	0.2454	0.1256	0.0553	0.0207
6	1.0000	0.9976	0.9781	0.9133	0.7858	0.6080	0.4166	0.2500	0.1299	0.0577
7	1.0000	0.9996	0.9941	0.9679	0.8982	0.7723	0.6010	0.4159	0.2520	0.1316
8	1.0000	0.9999	0.9987	0.9900	0.9591	0.8867	0.7624	0.5956	0.4143	0.2517
9	1.0000	1.0000	0.9998	0.9974	0.9861	0.9520	0.8782	0.7553	0.5914	0.4119
10	1.0000	1.0000	1.0000	0.9994	0.9961	0.9829	0.9468	0.8725	0.7507	0.5881

$p =$	0.05	0.10	0.15	0.20	0.25	0.30	0.35	0.40	0.45	0.50
11	1.0000	1.0000	1.0000	0.9999	0.9991	0.9949	0.9804	0.9435	0.8692	0.7483
12	1.0000	1.0000	1.0000	1.0000	0.9998	0.9987	0.9940	0.9790	0.9420	0.8684
13	1.0000	1.0000	1.0000	1.0000	1.0000	0.9997	0.9985	0.9935	0.9786	0.9423
14	1.0000	1.0000	1.0000	1.0000	1.0000	1.0000	0.9997	0.9984	0.9936	0.9793
15	1.0000	1.0000	1.0000	1.0000	1.0000	1.0000	1.0000	0.9997	0.9985	0.9941
16	1.0000	1.0000	1.0000	1.0000	1.0000	1.0000	1.0000	1.0000	0.9997	0.9987
17	1.0000	1.0000	1.0000	1.0000	1.0000	1.0000	1.0000	1.0000	1.0000	0.9998
18	1.0000	1.0000	1.0000	1.0000	1.0000	1.0000	1.0000	1.0000	1.0000	1.0000
$n = 25, x = 0$	0.2774	0.0718	0.0172	0.0038	0.0008	0.0001	0.0000	0.0000	0.0000	0.0000
1	0.6424	0.2712	0.0931	0.0274	0.0070	0.0016	0.0003	0.0001	0.0000	0.0000
2	0.8729	0.5371	0.2537	0.0982	0.0321	0.0090	0.0021	0.0004	0.0001	0.0000
3	0.9659	0.7636	0.4711	0.2340	0.0962	0.0332	0.0097	0.0024	0.0005	0.0001
4	0.9928	0.9020	0.6821	0.4207	0.2137	0.0905	0.0320	0.0095	0.0023	0.0005
5	0.9988	0.9666	0.8385	0.6167	0.3783	0.1935	0.0826	0.0294	0.0086	0.0020
6	0.9998	0.9905	0.9305	0.7800	0.5611	0.3407	0.1734	0.0736	0.0258	0.0073
7	1.0000	0.9977	0.9745	0.8909	0.7265	0.5118	0.3061	0.1536	0.0639	0.0216
8	1.0000	0.9995	0.9920	0.9532	0.8506	0.6769	0.4668	0.2735	0.1340	0.0539
9	1.0000	0.9999	0.9979	0.9827	0.9287	0.8106	0.6303	0.4246	0.2424	0.1148
10	1.0000	1.0000	0.9995	0.9944	0.9703	0.9022	0.7712	0.5858	0.3843	0.2122
11	1.0000	1.0000	0.9999	0.9985	0.9893	0.9558	0.8746	0.7323	0.5426	0.3450
12	1.0000	1.0000	1.0000	0.9996	0.9966	0.9825	0.9396	0.8462	0.6937	0.5000
13	1.0000	1.0000	1.0000	0.9999	0.9991	0.9940	0.9745	0.9222	0.8173	0.6550
14	1.0000	1.0000	1.0000	1.0000	0.9998	0.9982	0.9907	0.9656	0.9040	0.7878
15	1.0000	1.0000	1.0000	1.0000	1.0000	0.9995	0.9971	0.9868	0.9560	0.8852
16	1.0000	1.0000	1.0000	1.0000	1.0000	0.9999	0.9992	0.9957	0.9826	0.9461
17	1.0000	1.0000	1.0000	1.0000	1.0000	1.0000	0.9998	0.9988	0.9942	0.9784
18	1.0000	1.0000	1.0000	1.0000	1.0000	1.0000	1.0000	0.9997	0.9984	0.9927
19	1.0000	1.0000	1.0000	1.0000	1.0000	1.0000	1.0000	0.9999	0.9996	0.9980
20	1.0000	1.0000	1.0000	1.0000	1.0000	1.0000	1.0000	1.0000	0.9999	0.9995
21	1.0000	1.0000	1.0000	1.0000	1.0000	1.0000	1.0000	1.0000	1.0000	0.9999
22	1.0000	1.0000	1.0000	1.0000	1.0000	1.0000	1.0000	1.0000	1.0000	1.0000
$n = 30, x = 0$	0.2146	0.0424	0.0076	0.0012	0.0002	0.0000	0.0000	0.0000	0.0000	0.0000
1	0.5535	0.1837	0.0480	0.0105	0.0020	0.0003	0.0000	0.0000	0.0000	0.0000
2	0.8122	0.4114	0.1514	0.0442	0.0106	0.0021	0.0003	0.0000	0.0000	0.0000
3	0.9392	0.6474	0.3217	0.1227	0.0374	0.0093	0.0019	0.0003	0.0000	0.0000
4	0.9844	0.8245	0.5245	0.2552	0.0979	0.0302	0.0075	0.0015	0.0002	0.0000
5	0.9967	0.9268	0.7106	0.4275	0.2026	0.0766	0.0233	0.0057	0.0011	0.0002
6	0.9994	0.9742	0.8474	0.6070	0.3481	0.1595	0.0586	0.0172	0.0040	0.0007
7	0.9999	0.9922	0.9302	0.7608	0.5143	0.2814	0.1238	0.0435	0.0121	0.0026
8	1.0000	0.9980	0.9722	0.8713	0.6736	0.4315	0.2247	0.0940	0.0312	0.0081
9	1.0000	0.9995	0.9903	0.9389	0.8034	0.5888	0.3575	0.1763	0.0694	0.0214
10	1.0000	0.9999	0.9971	0.9744	0.8943	0.7304	0.5078	0.2915	0.1350	0.0494
11	1.0000	1.0000	0.9992	0.9905	0.9493	0.8407	0.6548	0.4311	0.2327	0.1002

$p =$	0.05	0.10	0.15	0.20	0.25	0.30	0.35	0.40	0.45	0.50
12	1.0000	1.0000	0.9998	0.9969	0.9784	0.9155	0.7802	0.5785	0.3592	0.1808
13	1.0000	1.0000	1.0000	0.9991	0.9918	0.9599	0.8737	0.7145	0.5025	0.2923
14	1.0000	1.0000	1.0000	0.9998	0.9973	0.9831	0.9348	0.8246	0.6448	0.4278
15	1.0000	1.0000	1.0000	0.9999	0.9992	0.9936	0.9699	0.9029	0.7691	0.5722
16	1.0000	1.0000	1.0000	1.0000	0.9998	0.9979	0.9876	0.9519	0.8644	0.7077
17	1.0000	1.0000	1.0000	1.0000	0.9999	0.9994	0.9955	0.9788	0.9286	0.8192
18	1.0000	1.0000	1.0000	1.0000	1.0000	0.9998	0.9986	0.9917	0.9666	0.8998
19	1.0000	1.0000	1.0000	1.0000	1.0000	1.0000	0.9996	0.9971	0.9862	0.9506
20	1.0000	1.0000	1.0000	1.0000	1.0000	1.0000	0.9999	0.9991	0.9950	0.9786
21	1.0000	1.0000	1.0000	1.0000	1.0000	1.0000	1.0000	0.9998	0.9984	0.9919
22	1.0000	1.0000	1.0000	1.0000	1.0000	1.0000	1.0000	1.0000	0.9996	0.9974
23	1.0000	1.0000	1.0000	1.0000	1.0000	1.0000	1.0000	1.0000	0.9999	0.9993
24	1.0000	1.0000	1.0000	1.0000	1.0000	1.0000	1.0000	1.0000	1.0000	0.9998
25	1.0000	1.0000	1.0000	1.0000	1.0000	1.0000	1.0000	1.0000	1.0000	1.0000
$n = 40, x = 0$	0.1285	0.0148	0.0015	0.0001	0.0000	0.0000	0.0000	0.0000	0.0000	0.0000
1	0.3991	0.0805	0.0121	0.0015	0.0001	0.0000	0.0000	0.0000	0.0000	0.0000
2	0.6767	0.2228	0.0486	0.0079	0.0010	0.0001	0.0000	0.0000	0.0000	0.0000
3	0.8619	0.4231	0.1302	0.0285	0.0047	0.0006	0.0001	0.0000	0.0000	0.0000
4	0.9520	0.6290	0.2633	0.0759	0.0160	0.0026	0.0003	0.0000	0.0000	0.0000
5	0.9861	0.7937	0.4325	0.1613	0.0433	0.0086	0.0013	0.0001	0.0000	0.0000
6	0.9966	0.9005	0.6067	0.2859	0.0962	0.0238	0.0044	0.0006	0.0001	0.0000
7	0.9993	0.9581	0.7559	0.4371	0.1820	0.0553	0.0124	0.0021	0.0002	0.0000
8	0.9999	0.9845	0.8646	0.5931	0.2998	0.1110	0.0303	0.0061	0.0009	0.0001
9	1.0000	0.9949	0.9328	0.7318	0.4395	0.1959	0.0644	0.0156	0.0027	0.0003
10	1.0000	0.9985	0.9701	0.8392	0.5839	0.3087	0.1215	0.0352	0.0074	0.0011
11	1.0000	0.9996	0.9880	0.9125	0.7151	0.4406	0.2053	0.0709	0.0179	0.0032
12	1.0000	0.9999	0.9957	0.9568	0.8209	0.5772	0.3143	0.1285	0.0386	0.0083
13	1.0000	1.0000	0.9986	0.9806	0.8968	0.7032	0.4408	0.2112	0.0751	0.0192
14	1.0000	1.0000	0.9996	0.9921	0.9456	0.8074	0.5721	0.3174	0.1326	0.0403
15	1.0000	1.0000	0.9999	0.9971	0.9738	0.8849	0.6946	0.4402	0.2142	0.0769
16	1.0000	1.0000	1.0000	0.9990	0.9884	0.9367	0.7978	0.5681	0.3185	0.1341
17	1.0000	1.0000	1.0000	0.9997	0.9953	0.9680	0.8761	0.6885	0.4391	0.2148
18	1.0000	1.0000	1.0000	0.9999	0.9983	0.9852	0.9301	0.7911	0.5651	0.3179
19	1.0000	1.0000	1.0000	1.0000	0.9994	0.9937	0.9637	0.8702	0.6844	0.4373
20	1.0000	1.0000	1.0000	1.0000	0.9998	0.9976	0.9827	0.9256	0.7870	0.5627
21	1.0000	1.0000	1.0000	1.0000	1.0000	0.9991	0.9925	0.9608	0.8669	0.6821
22	1.0000	1.0000	1.0000	1.0000	1.0000	0.9997	0.9970	0.9811	0.9233	0.7852
23	1.0000	1.0000	1.0000	1.0000	1.0000	0.9999	0.9989	0.9917	0.9595	0.8659
24	1.0000	1.0000	1.0000	1.0000	1.0000	1.0000	0.9996	0.9966	0.9804	0.9231
25	1.0000	1.0000	1.0000	1.0000	1.0000	1.0000	0.9999	0.9988	0.9914	0.9597
26	1.0000	1.0000	1.0000	1.0000	1.0000	1.0000	1.0000	0.9996	0.9966	0.9808
27	1.0000	1.0000	1.0000	1.0000	1.0000	1.0000	1.0000	0.9999	0.9988	0.9917
28	1.0000	1.0000	1.0000	1.0000	1.0000	1.0000	1.0000	1.0000	0.9996	0.9968

$p =$	0.05	0.10	0.15	0.20	0.25	0.30	0.35	0.40	0.45	0.50
29	1.0000	1.0000	1.0000	1.0000	1.0000	1.0000	1.0000	1.0000	0.9999	0.9989
30	1.0000	1.0000	1.0000	1.0000	1.0000	1.0000	1.0000	1.0000	1.0000	0.9997
31	1.0000	1.0000	1.0000	1.0000	1.0000	1.0000	1.0000	1.0000	1.0000	0.9999
32	1.0000	1.0000	1.0000	1.0000	1.0000	1.0000	1.0000	1.0000	1.0000	1.0000
$n = 50, x = 0$	0.0769	0.0052	0.0003	0.0000	0.0000	0.0000	0.0000	0.0000	0.0000	0.0000
1	0.2794	0.0338	0.0029	0.0002	0.0000	0.0000	0.0000	0.0000	0.0000	0.0000
2	0.5405	0.1117	0.0142	0.0013	0.0001	0.0000	0.0000	0.0000	0.0000	0.0000
3	0.7604	0.2503	0.0460	0.0057	0.0005	0.0000	0.0000	0.0000	0.0000	0.0000
4	0.8964	0.4312	0.1121	0.0185	0.0021	0.0002	0.0000	0.0000	0.0000	0.0000
5	0.9622	0.6161	0.2194	0.0480	0.0070	0.0007	0.0001	0.0000	0.0000	0.0000
6	0.9882	0.7702	0.3613	0.1034	0.0194	0.0025	0.0002	0.0000	0.0000	0.0000
7	0.9968	0.8779	0.5188	0.1904	0.0453	0.0073	0.0008	0.0001	0.0000	0.0000
8	0.9992	0.9421	0.6681	0.3073	0.0916	0.0183	0.0025	0.0002	0.0000	0.0000
9	0.9998	0.9755	0.7911	0.4437	0.1637	0.0402	0.0067	0.0008	0.0001	0.0000
10	1.0000	0.9906	0.8801	0.5836	0.2622	0.0789	0.0160	0.0022	0.0002	0.0000
11	1.0000	0.9968	0.9372	0.7107	0.3816	0.1390	0.0342	0.0057	0.0006	0.0000
12	1.0000	0.9990	0.9699	0.8139	0.5110	0.2229	0.0661	0.0133	0.0018	0.0002
13	1.0000	0.9997	0.9868	0.8894	0.6370	0.3279	0.1163	0.0280	0.0045	0.0005
14	1.0000	0.9999	0.9947	0.9393	0.7481	0.4468	0.1878	0.0540	0.0104	0.0013
15	1.0000	1.0000	0.9981	0.9692	0.8369	0.5692	0.2801	0.0955	0.0220	0.0033
16	1.0000	1.0000	0.9993	0.9856	0.9017	0.6839	0.3889	0.1561	0.0427	0.0077
17	1.0000	1.0000	0.9998	0.9937	0.9449	0.7822	0.5060	0.2369	0.0765	0.0164
18	1.0000	1.0000	0.9999	0.9975	0.9713	0.8594	0.6216	0.3356	0.1273	0.0325
19	1.0000	1.0000	1.0000	0.9991	0.9861	0.9152	0.7264	0.4465	0.1974	0.0595
20	1.0000	1.0000	1.0000	0.9997	0.9937	0.9522	0.8139	0.5610	0.2862	0.1013
21	1.0000	1.0000	1.0000	0.9999	0.9974	0.9749	0.8813	0.6701	0.3900	0.1611
22	1.0000	1.0000	1.0000	1.0000	0.9990	0.9877	0.9290	0.7660	0.5019	0.2399
23	1.0000	1.0000	1.0000	1.0000	0.9996	0.9944	0.9604	0.8438	0.6134	0.3359
24	1.0000	1.0000	1.0000	1.0000	0.9999	0.9976	0.9793	0.9022	0.7160	0.4439
25	1.0000	1.0000	1.0000	1.0000	1.0000	0.9991	0.9900	0.9427	0.8034	0.5561
26	1.0000	1.0000	1.0000	1.0000	1.0000	0.9997	0.9955	0.9686	0.8721	0.6641
27	1.0000	1.0000	1.0000	1.0000	1.0000	0.9999	0.9981	0.9840	0.9220	0.7601
28	1.0000	1.0000	1.0000	1.0000	1.0000	1.0000	0.9993	0.9924	0.9556	0.8389
29	1.0000	1.0000	1.0000	1.0000	1.0000	1.0000	0.9997	0.9966	0.9765	0.8987
30	1.0000	1.0000	1.0000	1.0000	1.0000	1.0000	0.9999	0.9986	0.9884	0.9405
31	1.0000	1.0000	1.0000	1.0000	1.0000	1.0000	1.0000	0.9995	0.9947	0.9675
32	1.0000	1.0000	1.0000	1.0000	1.0000	1.0000	1.0000	0.9998	0.9978	0.9836
33	1.0000	1.0000	1.0000	1.0000	1.0000	1.0000	1.0000	0.9999	0.9991	0.9923
34	1.0000	1.0000	1.0000	1.0000	1.0000	1.0000	1.0000	1.0000	0.9997	0.9967
35	1.0000	1.0000	1.0000	1.0000	1.0000	1.0000	1.0000	1.0000	0.9999	0.9987
36	1.0000	1.0000	1.0000	1.0000	1.0000	1.0000	1.0000	1.0000	1.0000	0.9995
37	1.0000	1.0000	1.0000	1.0000	1.0000	1.0000	1.0000	1.0000	1.0000	0.9998
38	1.0000	1.0000	1.0000	1.0000	1.0000	1.0000	1.0000	1.0000	1.0000	1.0000

THIS PAGE HAS INTENTIONALLY BEEN LEFT BLANK

Answers

TOPIC-BASED QUESTIONS

Pages 166–169: Algebra and Functions

1. $3 \times \left(\frac{1}{32} x^5\right)^{\frac{3}{5}} = 3 \times \left(\frac{1}{32}\right)^{\frac{3}{5}} \times x^{5 \times \frac{3}{5}} = 3 \times \left(\frac{1}{2^5}\right)^{\frac{3}{5}} \times x^3$ **[1 for x^3; 1 for 2^5]**

$= 3 \times \frac{1}{2^{5 \times \frac{3}{5}}} \times x^3 = 3 \times \frac{1}{2^3} \times x^3 = \frac{3}{8} x^3$ **[1]**

2. Simplify $\sqrt{8} = 2\sqrt{2}$ **[1]** and rearrange the denominator to be in the form $a + b\sqrt{c}$:

$\frac{2 + 4\sqrt{8}}{\sqrt{8} - 4} = \frac{2 + 8\sqrt{2}}{-4 + 2\sqrt{2}} \times \frac{-4 - 2\sqrt{2}}{-4 - 2\sqrt{2}} = \frac{-8 - 4\sqrt{2} - 32\sqrt{2} - 32}{16 + 8\sqrt{2} - 8\sqrt{2} - 8}$ **[1]**

$= \frac{-40 - 36\sqrt{2}}{8} = -5 - \frac{9\sqrt{2}}{2}$ **[1]**

$p = -5$ and $q = \frac{9}{2}$ **[1]**

3. $2x^3 - 4x^2 - 30x = 2x(x^2 - 2x - 15)$ **[1]**

$= 2x(x^2 + 3x - 5x - 15) = 2x((x + 3) - 5(x + 3))$

$= 2x(x + 3)(x - 5)$ **[1]**

> Remember to look for common factors of all terms first.

4. $(2x - 3y)(3x + 2y)(x - y) = (6x^2 + 4xy - 9xy - 6y^2)(x - y)$ **[1]**

$= (6x^3 - 6x^2y + 4x^2y - 4xy^2 - 9x^2y + 9xy^2 - 6xy^2 + 6y^3)$ **[1]**

$= 6x^3 - 11x^2y - xy^2 + 6y^3$

$a = 6$, $b = -11$, $c = -1$, $d = 6$ **[1]**

> Be careful with the signs here – watch out when multiplying by negative terms.

5. a) If $f(-2) = 0$, then $(x + 2)$ is a factor. **[1]**

$f(-2) = (2 \times (-2)^3) + (-2)^2 - (3 \times -2) + 6 = 0$ **[1]**

b) $(2x^3 + x^2 - 3x + 6) \div (x + 2)$
[1 for attempting division]

$= 2x^2 - 3x + 3$ **[1 for correct quotient]**

Then the discriminant $b^2 - 4ac = (-3)^2 - (4 \times 2 \times 3)$
$= -15 < 0$ **[1]**, so $x = 2$ is the only real root of $f(x) = 0$. **[1 for correct explanation]**

6. The first four terms in ascending powers of x are:

$\left(3 - \frac{x}{4}\right)^8 = \binom{8}{0} 3^8 \left(-\frac{x}{4}\right)^0 + \binom{8}{1} 3^7 \left(-\frac{x}{4}\right)^1 + \binom{8}{2} 3^6 \left(-\frac{x}{4}\right)^2 +$

$\binom{8}{3} 3^5 \left(-\frac{x}{4}\right)^3 = 6561 - 4374x + \frac{5103}{4} x^2 - \frac{1701}{8} x^3$

[5 marks: 1 for method; 1 for each correct term]

To find an approximation for 2.995, find the value of x for which $3 - \frac{x}{4} = 2.995$, $x = 0.02$ **[1]**

Substitute, $6561 - (4374 \times 0.02) + \frac{5103}{4}(0.02)^2 - \frac{1701}{8}(0.02)^3 = 6474.028599$ **[1]**

7. a) If $f(x) = 0$ has one real root, then $b^2 - 4ac = 0$.
For the given $f(x)$, $a = 9$, $b = -9k$ and $c = 3k$
$(-9k)^2 - (4 \times 9 \times 3k) = 81k^2 - 108k = 0$ **[1]**
$k(81k - 108) = 0$, $k = 0$ **[1]** or $81k - 108 = 0$, $k = \frac{4}{3}$ **[1]**

b) Given $k > 0$, $k = \frac{4}{3}$, the equation is thus
$f(x) = 9x^2 - 12x + 4$ **[1]** $= (3x - 2)^2$ **[1]**
$f(x) = 0$, $0 = (3x - 2)^2$
$3x - 2 = 0$
$x = \frac{2}{3}$ **[1]**

8. The discriminant shows that $y = -x^2 - 2x - 2$ has no real roots.
$b^2 - 4ac = (-2)^2 - (4 \times (-1) \times (-2)) = -4 < 0$ **[1]**

> Complete the square to find the turning point.

$y = -x^2 - 2x - 2 = -(x - 1)^2 - 1$ **[1 for method]**

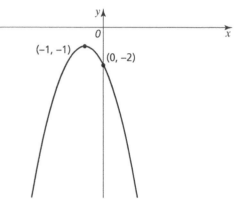

[1 for correct shape; 1 for turning point; 1 for y-intercept]

9. The curve and the line intersect when $f(x) = g(x)$
$-x^2 + 3x - 6 = -5x + 10$ **[1]**

> Rearrange and use the discriminant.

$-x^2 + 3x - 6 = -5x + 10$
$-x^2 + 8x - 16 = 0$
$b^2 - 4ac = 8^2 - (4 \times (-1) \times (-16)) = 0$ **[1]**
There is exactly one solution when the discriminant is equal to zero. **[1]**

> The point of intersection can be found by factorising, completing the square, or using the quadratic formula.

Using the quadratic formula (since discriminant already found):

$x = \frac{-8 \pm \sqrt{0}}{(2 \times -1)} = 4$ **[1]**

Substitute into $g(x) = (-5 \times 4) + 1 = -10$
The point of intersection is $(4, -10)$ **[1]**

> Don't forget to answer the question and write out the coordinates.
>
> NB: You could also have shown that the equation $-x^2 + 8x - 16 = -(x - 4)^2$, so $x = 4$ is a repeated root. Just be sure to explain each step in a 'show that' question.

10. Rearrange $2x + 3y = 8$

$x = 4 - \dfrac{3}{2}y$ **[1]** and substitute:

$\left(4 - \dfrac{3}{2}y\right)^2 + \left(2 \times \left(4 - \dfrac{3}{2}y\right) \times y\right) + y^2 = 16$

$\dfrac{9}{4}y^2 - 12y + 16 + 8y - 3y^2 + y^2 = 16$ **[1]**

$\dfrac{1}{4}y^2 - 4y = 0$

$y\left(\dfrac{1}{4}y - 4\right) = 0$ **[1]**

$y = 0$ **or** $\dfrac{1}{4}y - 4 = 0$, $y = 16$ **[1]**

Substituting:

When $y = 0$, $2x + (3 \times 0) = 8$, so $x = 4$

When $y = 16$, $2x + (3 \times 16) = 8$, so $x = -20$

The answers are $x = -20$, $y = 16$ and $x = 4$, $y = 0$ **[1]**

11. Sketch the graphs of $y < x^2 + 3x - 4$ and $y > \dfrac{1}{2}x - 3$.

The x-intercepts of the parabola are found by setting $y = 0$

$0 = x^2 + 3x - 4 = (x - 1)(x + 4)$

$x = 1$ and $x = 4$ **[1]**

The y-intercept is $y = -4$.

The line $y > \dfrac{1}{2}x - 3$ has gradient of $\dfrac{1}{2}$ and y-intercept of $y = -3$.

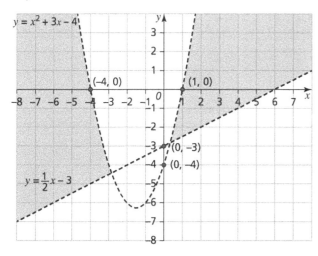

[2 for parabola; 1 for x-intercepts; 1 for y-intercept; 1 for line drawn correctly; 1 for dotted lines and correct shading]

Solve the simultaneous equations $y = x^2 + 3x - 4$ and $y = \dfrac{1}{2}x - 3$ to find the points of intersection.

$x^2 + 3x - 4 = \dfrac{1}{2}x - 3$

$x^2 + \dfrac{5}{2}x - 1 = 0$

Complete the square (or use the quadratic formula).

$\left(x + \dfrac{5}{4}\right)^2 = 1 + \left(\dfrac{5}{4}\right)^2$

$\left(x + \dfrac{5}{4}\right)^2 = \dfrac{41}{16}$ **[1]**

$\left(x + \dfrac{5}{4}\right)^2 = -\dfrac{5}{4} \pm \dfrac{\sqrt{41}}{4}$

$x = \dfrac{-5 + \sqrt{41}}{4}, x = \dfrac{-5 - \sqrt{41}}{4}$ **[1]**

The range of values of x for which $y < x^2 + 3x - 4$ and $y > \dfrac{1}{2}x - 3$ is $x > \dfrac{-5 + \sqrt{41}}{4}$ or $x < \dfrac{-5 - \sqrt{41}}{4}$

$\left\{x : x > \dfrac{-5 + \sqrt{41}}{4}\right\} \cup \left\{x : x < \dfrac{-5 - \sqrt{41}}{4}\right\}$ **[1]**

12. The coordinates of the point of intersection are when $f(x) = g(x)$

$2x^2 + 10x - 23 = -4x^2 - 2x + 25$

$6x^2 + 12x - 48 = 0$ **[1]**

$x^2 + 2x - 8 = 0$

$(x + 4)(x - 2) = 0$

$x = -4$ and $x = 2$ **[1]**

When $x = -4$, $2 \times (-4^2) + (10 \times -4) - 23 = -31$

When $x = 2$, $2 \times (2^2) + (10 \times 2) - 23 = 5$

The points of intersection are $(-4, -31)$ **[1]** and $(2, 5)$ **[1]**

Hence, the range of values of x for which $g(x) > f(x)$ is $-4 < x < 2$ **[1]**

13. a) If $x = 3$ is a root, then $f(3) = 0$

$f(3) = 2 \times (3^3) - 9 \times (3)^2 + (7 \times 3) + 6 = 0$ **[1]**

b) To sketch the curve,

$(2x^3 - 9x^2 + 7x + 6) \div (x - 3) = 2x^2 - 3x - 2$

[1 for method of division; 1 for correct quotient]

$2x^2 - 3x - 2 = (2x + 1)(x - 2)$ **[1]**

Then $f(x) = (x - 3)(2x + 1)(x - 2)$

The x-intercepts are when $f(x) = 0$, $x = 3$, $x = -\dfrac{1}{2}$ and $x = 2$.   **[1]**

The y-intercept is $y = 6$.

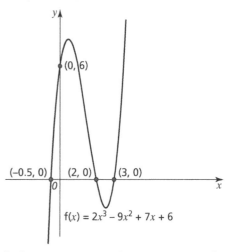

[1 for x-intercepts; 1 for y-intercept; 1 for general shape]

Answers

14. a)

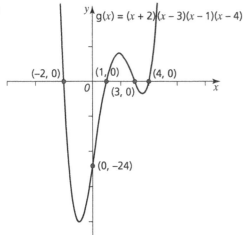

[1 for x-intercepts; 1 for y-intercept; 1 for general shape]

b) g(x – 1) is a translation of +1 in the x-direction so the points of intersection with the x-axis are (–1, 0), (2, 0), (4, 0) and (5, 0).

[2 for all correct; 1 for at least two correct]

The function g(x – 1) = (x – 1 + 2)(x – 1 – 3)(x – 1 – 1) (x – 1 – 4) = (x + 1)(x – 4)(x – 2)(x – 5)   **[1]**

The y-intercept is (0 + 1)(0 – 4)(0 – 2)(0 – 5) = –40 (0, –40)   **[1]**

15. $y = \dfrac{3}{(x-1)^2} + 2$ can be written as $y = f(x - 1) + 2$ **[1]**

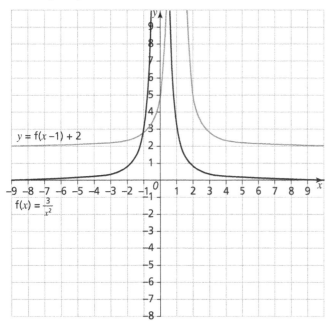

[2 marks: 1 for translation in x-direction; 1 for translation in y-direction]

$y = \dfrac{3}{(x-1)^2} + 2$ has asymptotes at $x = 1$ and $y = 2$ **[1]**

16. $y = $ f($-x$) is a reflection in the y-axis.

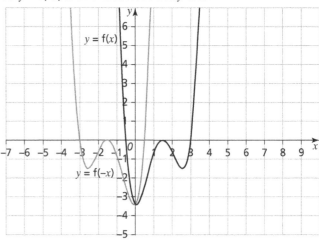

[1 for reflection in y-axis; 1 for general shape]

$y = -$f(x) is a reflection in the x-axis.

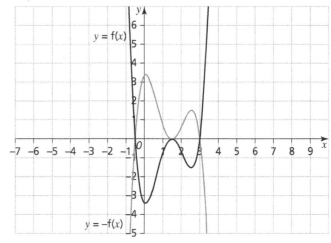

[1 for reflection; 1 for general shape]

Pages 170–171: Coordinate Geometry

1. Find the gradient, $m = \dfrac{y_2 - y_1}{x_2 - x_1} = \dfrac{5 - 8}{-2 - (-4)} = -\dfrac{3}{2}$ **[1]**

Substitute into $y - y_1 = m(x - x_1)$:

$y - 8 = -\dfrac{3}{2}(x - (-4))$ **[1]**

$y = -\dfrac{3}{2}x + 2$ **[1]**

2. a) $m_{AB} = \dfrac{y_2 - y_1}{x_2 - x_1} = \dfrac{3 - 2}{5 - 2} = \dfrac{1}{3}$ **[1]**

$m_{BC} = \dfrac{y_2 - y_1}{x_2 - x_1} = \dfrac{3 - 0}{5 - 6} = -3$ **[1]**; $-3 \times \dfrac{1}{3} = -1$ **[1]**, so AB and BC are perpendicular.

Other methods, such as using Pythagoras' Theorem, are also acceptable.

b) $d = \sqrt{(x_2 - x_1)^2 + (y_2 - y_1)^2}$

$d_{AB} = \sqrt{(5 - 2)^2 + (3 - 2)^2} = \sqrt{10}$ **[1]**

$d_{BC} = \sqrt{(6 - 5)^2 + (0 - 3)^2} = \sqrt{10}$ **[1]**

$A = \dfrac{bh}{2} = \dfrac{\sqrt{10} \times \sqrt{10}}{2} = 5$ units^2 **[1]**

3. $m_{AB} = \frac{0-4}{2-(-1)} = -\frac{4}{3}$ [1]

$y - y_1 = m(x - x_1)$

$y - 2 = -\frac{4}{3}(x - 3)$ [1]

$3y - 6 = -4(x - 3)$

$4x + 3y - 18 = 0$ [1]

4. $m_{AB} = \frac{2-5}{4-1} = -1$ **[1]**; the gradient of the perpendicular bisector will be 1 because $1 \times -1 = -1$ **[1]**

The midpoint of AB is

$\left(\frac{x_1+x_2}{2}, \frac{y_1+y_2}{2}\right) = \left(\frac{1+4}{2}, \frac{5+2}{2}\right) = \left(\frac{5}{2}, \frac{7}{2}\right)$ [1]

Substitute into $y - y_1 = m(x - x_1)$

$y - \frac{7}{2} = -1\left(x - \frac{5}{2}\right)$ [1]

$y = -x + 6$ [1]

5. Perpendicular bisector of AB:

$m_{AB} = -\frac{8-4}{5-(-3)} = \frac{1}{2}$ **[1 for both m_{AB} and m_{AC}]**

Midpoint of AB $= \left(\frac{-3+5}{2}, \frac{4+8}{2}\right) = (1, 6)$

[1 for both midpoint of AB and of AC]

Equation of the perpendicular bisector of AB:

$y - 6 = -2(x - 1)$

$y = -2x + 8$

$2x + y - 8 = 0$ [1]

> Rearranging this equation sets it up for solving the simultaneous equations.

Perpendicular bisector of AC:

$m_{AC} = -\frac{1-4}{-2-(-3)} = -3$

Midpoint of AC $= \left(\frac{-3-2}{2}, \frac{4+1}{2}\right) = \left(-\frac{5}{2}, \frac{5}{2}\right)$

Equation of the perpendicular bisector of AC:

$y - \frac{5}{2} = \frac{1}{3}\left(x + \frac{5}{2}\right)$

$6y - 15 = 2x + 5$

$-2x + 6y - 20 = 0$ [1]

Simultaneous equations:

$2x + y - 8 = 0$

$-2x + 6y - 20 = 0$

[1 for setting up simultaneous equations]

$7y - 28 = 0$

$y = 4$

$4 = -2x + 8$, so $x = 2$

Centre (2, 4) **[1 for both x and y value correct]**

$r = d = \sqrt{(-3-2)^2 + (4-4)^2} = 5$ [1]

Equation of the circle: $(x-2)^2 + (y-4)^2 = 25$ [1]

6. a) $(x - 4k)^2 + (y - k)^2 = 10$

$(7 - 4k)^2 + (0 - k)^2 = 10$ [1]

$17k^2 - 56k + 39 = 0$

$(k - 1)(17k - 39) = 0$ [1]

$k = 1$ or $k = \frac{39}{17}$ [1]

b) $(x - 4)^2 + (y - 1)^2 = 10$, so the centre is (4, 1) [1]

7. a) Complete the square:

$x^2 - 4x + y^2 - 6y = 2$

$(x - 2)^2 + (y - 3)^2 = 2 + 4 + 9$ [1]

$(x - 2)^2 + (y - 3)^2 = 15$

Centre (2, 3) **[1]**; radius $\sqrt{15}$ **[1]**

b) The x-intercepts are when $y = 0$

$x^2 - 4x + (0)^2 - (6 \times 0) = 2$ [1]

$x^2 - 4x = 2$ **[1 for method of either completing the square or using the quadratic formula]**

$(x - 2)^2 = 6$

$x = 2 + \sqrt{6}$ and $x = 2 - \sqrt{6}$

The coordinates are $\left(2 - \sqrt{6}, 0\right)$ and $\left(2 + \sqrt{6}, 0\right)$ [1]

Pages 172–173: Trigonometry

1. a) $a^2 = 19^2 + 8^2 - 2 \times 8 \times 19 \times \cos 127$ [1]

$a^2 = 607.95$ [1]

$a = 24.7\,\text{cm}$ [1]

b) $\frac{19}{\sin B} = \frac{24.7}{\sin 127}$ [1]

$\sin B = 0.614$ [1]

$B = 38°$ (2 s.f.) [1]

c) $180 - 38 - 127$ [1]

$15°$ (2 s.f.) [1]

2. $\tan x = -\frac{1}{3}$ [2]

$\tan^{-1}\left(-\frac{1}{3}\right) = -18.4$ (not in range) [1]

$161.6, 341.6$ [1]

3. a) $1 + 2\sin x + \sin^2 x + \cos^2 x$ [1]

$1 + 2\sin x + 1$ [1]

$2 + 2\sin x$ [1]

b) $2 + 2\sin x = 1$ [1]

$\sin x = -\frac{1}{2}$ [1]

$x = \frac{-\pi}{6}$ (not in range) [1]

$x = \frac{7\pi}{6}, \frac{11\pi}{6}$ [1]

4. a)

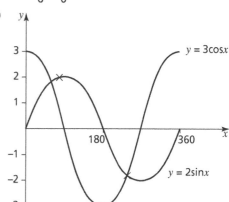

Answers

[1 for correct shapes; 1 for sin graph between –2 and 2; 1 for cos graph between –3 and 3]

b) 2 [1]

Pages 174–175: Exponentials and Logarithms

1. $\log_2(3x - 1) = 4$

$3x - 1 = 2^4 = 16$ [2]

$x = \dfrac{17}{3}$ [1]

2. $\ln(5 - 2x) = 10$

$5 - 2x = e^{10}$ [2]

$x = \dfrac{5 - e^{10}}{2}$ [1]

3. $\log_{10}(x + 3) - \log_{10}(2x - 1) = 3$

$\log_{10}\left(\dfrac{x + 3}{2x - 1}\right) = 3$ [2]

$\dfrac{x + 3}{2x - 1} = 10^3 = 1000$ [1]

$x + 3 = 2000x - 1000$

$1003 = 1999x$

$x = \dfrac{1003}{1999}$ [1]

4. $3^{1-x} = 2^x$

$(1 - x)\log_{10}3 = x\log_{10}2$ [2]

$\log_{10}3 - x\log_{10}3 = x\log_{10}2$

$\log_{10}3 = x(\log_{10}2 + \log_{10}3)$

$x = \dfrac{(\log_{10}2 + \log_{10}3)}{\log_{10}3} = 1.63$ [2]

5. $y = -\ln(x + 1)$

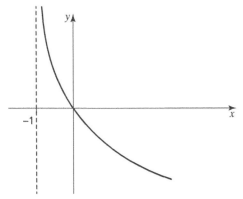

[1 for shape; 1 for asymptote at $x = -1$; 1 for intercept at (0, 0)]

6. $e^{2x} - 3e^x - 4 = 0$

$(e^x - 4)(e^x + 1) = 0$ [2]

$e^x = 4$ or $e^x = -1$ [2]

Only solution is $x = \ln4$ [1]

7. $4\log_4 2 + \log_4 8 - \log_4 2$

$= \log_4 16 + \log_4 8 - \log_4 2$ [1]

$= \log_4 16 + \log_4 4$ [1]

$= 2 + 1 = 3$ [1]

8. $2\log_2(x - 9) = \log_3 81 + 1$

$2\log_2(x - 9) = 4 + 1$ [1]

$\log_2(x - 9) = \dfrac{5}{2}$ [1]

$x - 9 = 2^{\frac{5}{2}}$ [2]

$x - 9 = \sqrt{32}$

$x = 9 + 4\sqrt{2}$ only solution [1]

9. $\ln\left(\dfrac{p^5}{q^6}\right)^{\frac{3}{2}} = \ln\left(\dfrac{p^{\frac{15}{2}}}{q^9}\right)$ [1]

$= \ln p^{\frac{15}{2}} - \ln q^9$ [1]

$= \dfrac{15}{2}\ln p - 9\ln q$ [1]

10. $\log_6 z = 4\log_z 2 + 4\log_z 3$

$\log_6 z = 4(\log_z 2 + \log_z 3)$

$\log_6 z = 4\log_z 6$ [1]

$\log_6 z = \dfrac{4}{\log_6 z}$ Use the change of base formula. [1]

$(\log_6 z)^2 = 4$ [1]

$\log_6 z = 2$ or $\log_6 z = -2$ Take square roots. [2]

So $z = 6^2 = 36$ or $z = 6^{-2} = \dfrac{1}{36}$ [2]

Pages 176–177: Differentiation

1. a) $\dfrac{dy}{dx} = x^3 + x^2 - 12x$ [1]

$x(x^2 + x - 12) = 0$ [1]

$x(x + 4)(x - 3), x = 0, x = -4, x = 3$ [1]

$\dfrac{d^2y}{dx^2} = 3x^2 + 2x - 12$ [1]

At $x = 0$, $\dfrac{d^2y}{dx^2} = $ –ve maximum

At $x = -4$, $\dfrac{d^2y}{dx^2} = $ +ve minimum

At $x = 3$, $\dfrac{d^2y}{dx^2} = $ +ve minimum

[1 for all three conclusions]

b)

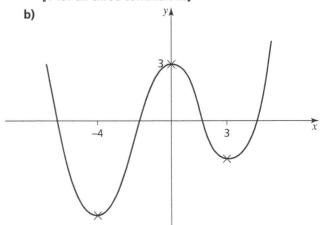

[1 for correct shape; 1 for minimum at $x = -4$ and $x = 3$; 1 for maximum at (0, 3)]

2. $3x^3 - x^{\frac{1}{2}} + \dfrac{x^3}{x^3} + \dfrac{4x}{x^3}$ **[1]**

$3x^3 - x^{\frac{1}{2}} + 1 + 4x^{-2}$ **[1]**

$9x^2 - \dfrac{1}{2}x^{-\frac{1}{2}} - 8x^{-3}$ **[2]**

3. a) $SA = 2\pi r^2 + 2\pi rh$ **[1]**

$\pi r^2 h = 200$ **[1]**

$h = \dfrac{200}{\pi r^2}$ **[1]**

$SA = 2\pi r^2 + 2\pi r \times \dfrac{200}{\pi r^2}$

$SA = 2\pi r^2 + \dfrac{400}{r}$ **[1]**

b) $\dfrac{\mathrm{d}(SA)}{\mathrm{d}r} = 4\pi r - 400r^{-2}$ **[1]**

$4\pi r - 400r^{-2} = 0$ **[1]**

$r^3 = \dfrac{400}{4\pi}$, $r = 3.17$ **[1]**; $SA = 189.3$ **[1]**

4. $\dfrac{\mathrm{d}y}{\mathrm{d}x} = 2x - 7$ **[1]**

at $x = 2$, $\dfrac{\mathrm{d}y}{\mathrm{d}x} = -3$ **[1]**

$y - 0 = -3(x - 2)$ **[1]**

$y = -3x + 6$ **[1]**

Pages 178–179: Integration

1. $x - x^2 = 0$, $x(x - 1) = 0$ **[1]**; $x = 0$ and $x = 1$ **[1]**

$\displaystyle\int_0^1 x - x^2 \mathrm{d}x$ **[1]**

$\left[\dfrac{x^2}{2} - \dfrac{x^3}{3}\right]_0^1$ **[1]**

$\left(\dfrac{1^2}{2} - \dfrac{1^3}{3}\right) - \left(\dfrac{0^2}{2} - \dfrac{0^3}{3}\right) = \dfrac{1}{6}$ **[1]**

2. a) $x^{\frac{5}{2}} + 2x^2 - x^{\frac{3}{2}} - 2x$ **[2]**

$\displaystyle\int = \dfrac{2}{7}x^{\frac{7}{2}} + \dfrac{2}{3}x^3 - \dfrac{2}{5}x^{\frac{5}{2}} - x^2 + c$ **[2]**

b) $\left(\dfrac{2}{7}(3)^{\frac{7}{2}} + \dfrac{2}{3}(3)^{\frac{3}{2}} - \dfrac{2}{5}(3)^{\frac{5}{2}} - (3)^2\right) -$

$\dfrac{2}{7}(1)^{\frac{7}{2}} + \dfrac{2}{3}(1)^{\frac{3}{2}} - \dfrac{2}{5}(1)^{\frac{5}{2}} - (1)^2$ **[1]**

16.6 (3 s.f.) **[1]**

3. a) $y = (3x + 5)^3$ **[2; 1 for attempt to cube both sides]**

b) $\displaystyle\int (3x + 5)^3 \, \mathrm{d}x$ **[1]**

$\displaystyle\int 27x^3 + 135x^2 + 225x + 125 \mathrm{d}x$ **[1]**

$\dfrac{27x^4}{4} + 45x^3 + \dfrac{225x^2}{2} + 125x + c$ **[1]**

4. $\displaystyle\int 3x^2 + 4x - 3x^{\frac{-1}{2}} \, \mathrm{d}x$ **[1]**

$y = x^3 + 2x^2 - 6x^{\frac{1}{2}} + c$ **[1]**

$4 = 1^3 + 2(1)^2 - 6(1)^{\frac{1}{2}} + c$ **[1]**

$c = 7$ **[1]**

$y = x^3 + 2x^2 - 6x^{\frac{1}{2}} + 7$ **[1]**

Pages 180–181: Vectors

1. Let C be the required point.

$\overrightarrow{AB} = \overrightarrow{OB} - \overrightarrow{OA} = \begin{pmatrix} 12 \\ 26 \end{pmatrix} - \begin{pmatrix} 0 \\ 10 \end{pmatrix} = \begin{pmatrix} 12 \\ 16 \end{pmatrix}$ **[2]**

$\overrightarrow{AC} = \dfrac{1}{5}\overrightarrow{AB} = \begin{pmatrix} \frac{12}{5} \\ \frac{16}{5} \end{pmatrix}$ **[2]**

$\overrightarrow{OC} = \overrightarrow{OA} + \overrightarrow{AC} = \begin{pmatrix} 0 \\ 10 \end{pmatrix} + \begin{pmatrix} \frac{12}{5} \\ \frac{16}{5} \end{pmatrix}$ **[1]**

$= \begin{pmatrix} \frac{12}{5} \\ \frac{66}{5} \end{pmatrix}$ **[1]**

2. Using distance formula:

$d = \sqrt{(2 + 5a)^2 + (-3a - 4)^2}$ **[2]**

$d^2 = 25a^2 + 20a + 4 + 9a^2 + 24a + 16$

$d^2 = 34a^2 + 44a + 20$ **[1]**

$d^2 = 34\left(a^2 + \dfrac{44}{34}a + \dfrac{20}{34}\right)$ **[1]**

$d^2 = 34\left(\left(a + \dfrac{11}{17}\right)^2 - \left(\dfrac{11}{17}\right)^2 + \dfrac{20}{34}\right)$ **[1]**

$d^2 = 34\left(a + \dfrac{11}{17}\right)^2 + \dfrac{98}{17}$ **[1]**

Which is minimum when $a = -\dfrac{11}{17}$

3. $\overrightarrow{BC} = \begin{pmatrix} -1 \\ 2 \end{pmatrix}$ **[1]**

$\overrightarrow{OD} = \overrightarrow{OA} + \overrightarrow{AD} = \overrightarrow{OA} + \overrightarrow{BC}$

$= \begin{pmatrix} -2 \\ -8 \end{pmatrix} + \begin{pmatrix} -1 \\ 2 \end{pmatrix}$ **[1]**

$= \begin{pmatrix} -3 \\ -6 \end{pmatrix}$ **[1]**

Coordinates are D(−3, −6) **[1]**

4. $\mathbf{a} = 24\cos120\mathbf{i} + 24\sin120\mathbf{j}$ **[1]**

$= -12\mathbf{i} + 12\sqrt{3}\mathbf{j}$ **[1]**

5. Unit vector is $\dfrac{-\mathbf{i} + 3\mathbf{j}}{\sqrt{1^2 + 3^2}}$ **[1]**

$= \dfrac{-\mathbf{i} + 3\mathbf{j}}{\sqrt{10}}$ **[1]**

So require $\dfrac{2\sqrt{5}(-\mathbf{i} + 3\mathbf{j})}{\sqrt{10}}$ **[1]**

$\sqrt{2}(-\mathbf{i} + 3\mathbf{j})$ **[1]**

Answers

6. $\overrightarrow{OD} = \frac{1}{4}\mathbf{a}$ **[1]**

$\overrightarrow{OE} = \overrightarrow{OD} + \overrightarrow{DE}$ **[1]**

$= \frac{1}{4}\mathbf{a} + \lambda\left(-\frac{1}{4}\mathbf{a} + \mathbf{b}\right)$ **[1]**

$= \left(\frac{1}{4} - \frac{\lambda}{4}\right)\mathbf{a} + \lambda\mathbf{b}$

$\overrightarrow{OE} = \mu\overrightarrow{OC} = \mu\left(\overrightarrow{OA} + \overrightarrow{AC}\right)$ **[1]**

$= \mu\left(\mathbf{a} + \frac{1}{4}(-\mathbf{a} + \mathbf{b})\right)$ **[1]**

$= \frac{3\mu}{4}\mathbf{a} + \frac{\mu}{4}\mathbf{b}$

Equate **b**: $\lambda = \frac{\mu}{4}$ **[2]**

Equate **a**: $\frac{3\mu}{4} = \frac{1}{4} - \frac{\lambda}{4}$ **[2]**

Solve simultaneously to give $\lambda = \frac{1}{13}$ **[1]**

$\Rightarrow \overrightarrow{OE} = \frac{3}{13}\mathbf{a} + \frac{1}{13}\mathbf{b}$ **[1]**

7. Parallel $\Rightarrow \mathbf{p} = k\mathbf{q}$ **[1]**

$3\mathbf{a} + (2\lambda + 1)\mathbf{b} = k((2 - 5\lambda)\mathbf{a} - 12\mathbf{b}) = (2 - 5\lambda)k\mathbf{a} - 12k\mathbf{b}$

Equate **a**: $3 = (2 - 5\lambda)k$ **[1]**

Equate **b**: $2\lambda + 1 = -12k$ **[1]**

Solving simultaneously: **[1]**

$\frac{3}{2 - 5\lambda} = \frac{2\lambda + 1}{-12}$

$\Rightarrow \lambda = -2$ **[1]**

8. $|\mathbf{a}| = \sqrt{2^2 + 1^2} = \sqrt{5}$ **[2]**

Direction $= \tan^{-1}\left(-\frac{1}{2}\right) = -26.6°$, or 26.6° below x-axis. **[2]**

9. $|\mathbf{R}|^2 = 14^2 + 8^2 - 2 \times 14 \times 8 \times \cos 150°$ **[2]**

$\Rightarrow |\mathbf{R}| = 21.3$ **[1]**

$\frac{\sin \theta}{8} = \frac{\sin 150}{21.3}$ **[2]**

$\Rightarrow \theta = 10.8°$ **[1]**

Page 182: Proof

1. $(2k)^2 - 4 \times 1 \times -1 = 0$ **[1]**

Use $b^2 - 4ac = 0$

$4k^2 - 4 = 0$ **[1]**

$k = 1$ or -1 **[1]**

2. $2x^3 + x^2 - 18x - 9$ has a factor of $(x - 3)$

$2(3)^3 + (3)^2 - 18(3) - 9 = 0$ therefore factor **[2]**

$(2x^3 + x^2 - 18x - 9) \div (x - 3) = (2x^2 + 7x + 3)$ **[2]**

$2x^3 + x^2 - 18x - 9 = (x - 3)(2x + 1)(x + 3)$ **[1]**

3. $\left(y - \frac{5}{2}\right)^2 - \frac{25}{4} + 30$ **[1]**

$\left(y - \frac{5}{2}\right)^2 + \frac{95}{4}$

$\left(y - \frac{5}{2}\right)^2 > 0$ **[1]**

$\therefore \left(y - \frac{5}{2}\right)^2 + \frac{95}{4} > 0$ **[1]**

Page 183: Statistical Sampling & Statistical Hypothesis Testing

1. a) $P(X \leq 6) = 0.0238$

$P(X \leq 7) = 0.0553$

As $0.0238 < 0.025$ but $0.0553 > 0.025$ **[1]**

$1 - P(X \leq 17) = P(X \geq 18) = 1 - 0.9680 = 0.032$

$1 - P(X \leq 18) = P(X \geq 19) = 1 - 0.9852 = 0.0148$

As $0.032 > 0.025$ but $0.0148 < 0.025$ **[1]**

$X \geq 19$ is the upper critical region

$X \leq 6$ is the lower critical region

[1 for both correct regions]

b) $0.0148 + 0.0238 = 0.0386 = 3.86\%$ **[1]**

c) 5 lies within the critical region **[1]**

Therefore there is evidence to suggest X is not modelled by the binomial distribution B(40, 0.3). **[1]**

2. a) 0.0081 **[1]**

b) $P(X \geq 21) = 1 - P(X \leq 20)$ **[1]**

$1 - 0.9786 = 0.0214$ **[1]**

Pages 184–187: Data Presentation and Interpretation

1. a) $\frac{20}{2} + 0.5 = 10.5$th value **[1]**

9.35 **[1]**

b) $\frac{20}{4} + 0.5 = 5.5$th value   7.45

$5.5 \times 3 = 16.5$th value   9.95

IQR $= 9.95 - 7.45 = 2.5$ **[1]**

$9.95 + 1.5(2.5) = 13.7$ (upper boundary)

$7.45 - 1.5(2.5) = 3.7$ (lower boundary) **[1]**

No outliers **[1]**

c)

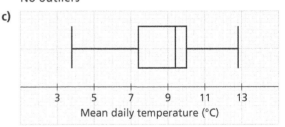

Mean daily temperature (°C)

[1 for correct box; 1 for correct median; 1 for correct whiskers]

d)

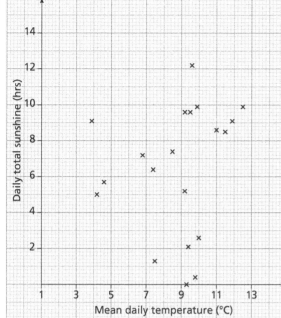

[2 for fully correct diagram; 1 if only one point incorrect]

The scatter diagram shows little correlation. **[1]**

This does not support Daniel's claim. **[1]**

2. Motorway B has a higher average speed. **[1]**

The median of motorway B is higher. **[1]**

Both motorways have a similar spread of speeds. **[1]**

The IQR for both motorways is very similar. **[1]**

3. a) Midpoints 132, 144.5, 152, 162 **[1]**

$\dfrac{9776.5}{67} = 145.9$ **[2]**

b) $\dfrac{1\,432\,294}{67} - \left(\dfrac{9776.5}{67}\right)^2 = 90.7$ **[2]**

$\sqrt{90.7} = 9.52$ **[1]**

c) Continuous data **[1]**

d)

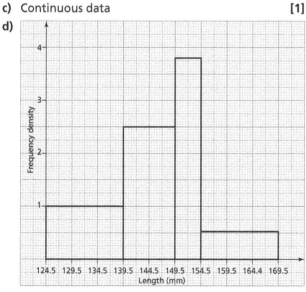

[2 for frequency densities of 1, 2.5, 3.8, 0.53; 2 for fully correct graph (1 if only one bar incorrect)]

Pages 188–189: Probability

1. a)

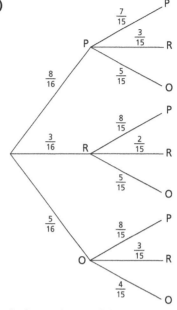

[1 for each set of three branches]

b) $\dfrac{5}{16} \times \dfrac{4}{15} = \dfrac{1}{12}$ **[2]**

c) $\dfrac{5}{16} \times \dfrac{4}{15} + \dfrac{8}{16} \times \dfrac{7}{15} + \dfrac{3}{16} \times \dfrac{2}{15}$ **[2]**

$\dfrac{1}{12} + \dfrac{7}{30} + \dfrac{1}{40} = \dfrac{41}{120}$ **[2]**

2. a) $0.2 + 0.3 + 0.15 = 0.65$ **[1]**

$1 - 0.65$ **[1]** $= 0.35$ **[1]**

b) $0.3 + 0.35 = 0.65$ **[1]**

$\dfrac{0.35}{0.65}$ **[1]** $= \dfrac{7}{11}$ **[1]**

c) P(carrots) = 0.55 **[1]**

P(peas) = 0.65 **[1]**

$0.55 \times 0.65 = 0.3575$ **[1]**

0.3575 is not 0.35, therefore not independent **[1]**

Page 190: The Language of Mechanics

1. $m = \dfrac{F}{a} = \dfrac{150}{2.5} = 60\,\text{kg}$ **[2]**

2. Displacement $= -18\,\text{m}$ **[1]**

Velocity $= \dfrac{\text{displacement}}{\text{time}} = -\dfrac{18}{12} = -1.5\,\text{ms}^{-1}$ **[2]**

3. Newton 2: $26 - R = 20 \times 0.6$ **[2]**

$R = 26 - 20 \times 0.6 = 14\,\text{N}$ **[1]**

4. Newton 2: Acceleration $= \dfrac{\text{force}}{\text{mass}} = \dfrac{20}{8} = 2.5\,\text{ms}^{-2}$ **[2]**

Change in velocity = acceleration × time

$= 2.5 \times 10 = 25\,\text{ms}^{-1}$ **[2]**

5. Acceleration $= \dfrac{(2.5 - 0.5)}{60} = \dfrac{1}{30}\,\text{ms}^{-2}$ **[2]**

Newton 2: $F = ma = 12 \times \dfrac{1}{30} = 0.4\,\text{N}$ **[2]**

Answers

Pages 191–193: Kinematics

1. a) $\frac{15}{2} = 7.5\ \text{ms}^{-2}$ [1]

   b) $\frac{15}{6} = 2.5\ \text{ms}^{-2}$ [1]

   c) $\frac{1}{2}(16 + 8)15 = 180\ \text{m}$ [2]

   d) $\frac{180}{16} = 11.25\ \text{ms}^{-1}$ [2]

2. a)

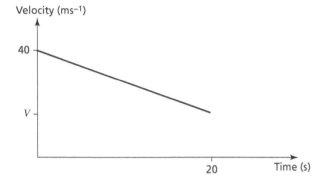

Velocity (ms⁻¹)

[1 for shape; 3 for 40, V, 20 marked]

   b) $\frac{1}{2}(40 + V)20 = 550$ [1]

       $V = 15\ \text{ms}^{-1}$ [1]

   c) $\frac{40 + 15}{2} = 27.5\ \text{ms}^{-1}$ [2]

3. a)

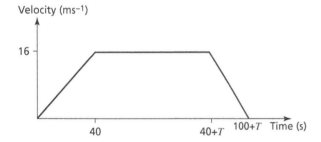

Velocity (ms⁻¹)

[1 for shape; 4 for 16, 40, 40 + T, 100 + T marked]

   b) $\frac{1}{2} \times 40 \times 16 + 16T + \frac{1}{2} \times 60 \times 16 = 1920$ [2]

       $\Rightarrow T = 70$ seconds   [1]

4. $s = ut + \frac{1}{2}at^2$

   $30 = 10t - \frac{t^2}{8}$ [2]

   $t^2 - 80t + 240 = 0$

   Using quadratic formula   [1]

   $t = 3.12\,\text{s}$ and $t = 76.9\,\text{s}$ [2]

5. a) $v^2 = u^2 + 2as \Rightarrow s = \frac{v^2 - u^2}{2a} = \frac{0^2 - 15^2}{2 \times (-3.5)} = 32.1\,\text{m}$ [2]

   b) $t = \frac{v - u}{a} = \frac{0 - 15}{-3.5} = 4.29$ seconds   [2]

6. $v = \frac{\text{d}s}{\text{d}t} = 2t^3 - 17t^2 + 35t$ [1]

   $2t^3 - 17t^2 + 35t = 0$ [1]

   $t(2t^2 - 17t + 35) = 0$ [1]

   $t(2t - 7)(t - 5) = 0$ [1]

   So $t = \frac{7}{2}$ [1]

   $s = \frac{\left(\frac{7}{2}\right)^4}{2} - \frac{17\left(\frac{7}{2}\right)^3}{3} + \frac{35\left(\frac{7}{2}\right)^2}{2}$

   $= 46.4\,\text{m}$ [2]

7. $v = \int a\text{d}t = \int\left(2t^2 - 1\right)\text{d}t = \frac{2t^3}{3} - t + c$ [2]

   $t = 3,\ v = 6 \Rightarrow 6 = 15 + c \Rightarrow c = -9$ [2]

   $v = \frac{2t^3}{3} - t - 9$

   $t = 4 \Rightarrow v = \frac{2 \times 4^3}{3} - 4 - 9 = 29.7$ seconds   [1]

8. $v > 0$ for $0 < t < 1$ and $v < 0$ for $1 < t < 2$

   $\int_0^1 (5 - 5t)\text{d}t = \left[5t - \frac{5t^2}{2}\right]_0^1 = 5 - \frac{5}{2} = \frac{5}{2}$ [2]

   $\int_1^2 (5 - 5t)\text{d}t = \left[5t - \frac{5t^2}{2}\right]_1^2 = -\frac{5}{2}$ [2]

   So distance travelled $= \frac{5}{2} + \frac{5}{2} = 5\,\text{m}$ [1]

9. $v = \int a\text{d}t = \int 2t\text{d}t = t^2 + c$ [1]

   $s = \int v\text{d}t = \int\left(t^2 + c\right)\text{d}t = \frac{t^3}{3} + ct + d$ [1]

   $t = 3 \Rightarrow s = 9 + 3c + d = 10$ [1]

   $t = 6 \Rightarrow s = 72 + 6c + d = 70$ [1]

   $3c + d = 1$

   $6c + d = -2$

   Solve simultaneously: [1]

   $c = -1,\ d = 4$ [2]

   $s = \frac{t^3}{3} - t + 4$

Pages 194–196: Forces and Newton's Laws

1. $v^2 = u^2 + 2as$

   $0 = 40^2 - 2gs$ [2]

   $s = \frac{40^2}{2g} = 81.6\,\text{m}$ [1]

2. $\mathbf{a} = \frac{(4\mathbf{i} + 8\mathbf{j}) - (-3\mathbf{i} - 9\mathbf{j})}{2} = \frac{7}{2}\mathbf{i} + \frac{17}{2}\mathbf{j}$ [2]

   $\mathbf{F} = m\mathbf{a} = 4\left(\frac{7}{2}\mathbf{i} + \frac{17}{2}\mathbf{j}\right) = 14\mathbf{i} + 34\mathbf{j}$ [2]

3. Newton 2   car: $1100 - T - 1200k = 1200 \times 0.5$ [2]

   Newton 2   caravan: $T - 800k = 800 \times 0.5$ [2]

   Adding equations: [1]

   $1100 - 2000k = 1000$

$k = \frac{1}{20}$ [1]

$\Rightarrow T = 400 + 800\left(\frac{1}{20}\right) = 440\,\text{N}$ [2]

4. Top particle: $s = \frac{1}{2}gt^2$ [2]

Bottom particle: $100 - s = 40t - \frac{g}{2}t^2$ [2]

Solving simultaneously: [1]

$100 - \frac{g}{2}t^2 = 40t - \frac{g}{2}t^2$

$\Rightarrow t = \frac{5}{2}$ [1]

So $100 - s = 100 - \frac{g}{2}\left(\frac{5}{2}\right)^2 = 69.4\,\text{m}$ from foot of cliff [2]

5. Equilibrium $\Rightarrow \mathbf{p} + \mathbf{q} + \mathbf{r} = 0$ [1]

$\begin{pmatrix} a \\ 2b \end{pmatrix} + \begin{pmatrix} b \\ 10 \end{pmatrix} + \begin{pmatrix} 3 \\ a \end{pmatrix} = \begin{pmatrix} 0 \\ 0 \end{pmatrix}$

$a + b + 3 = 0$ [1]

$2b + 10 + a = 0$ [1]

So $a + b = -3$

And $a + 2b = -10$

Solve simultaneously: [1]

$a = 4, b = -7$ [2]

6. Newton 2 to 3 kg mass: $18 - T_1 = 3a$ [2]

Newton 2 to 2 kg mass: $T_1 - T_2 = 2a$ [2]

Newton 2 to 1 kg mass: $T_2 = a$ [1]

Adding equations $\Rightarrow 18 = 6a \Rightarrow a = 3\,\text{ms}^{-2}$ [1]

> Can also be obtained by applying Newton 2 to the whole system.

$\Rightarrow T_2 = 3\,\text{N}$ [1]

$T_1 = T_2 + 2a = 3 + 2 \times 3 = 9\,\text{N}$ [1]

7. a) Newton 2   A: $T - 5mg = 5ma$ [2]

Newton 2   B: $6mg - T = 6ma$ [2]

Adding equations: [1]

$mg = 11ma$

$\Rightarrow a = \frac{g}{11}\,\text{ms}^{-2}$ [1]

b) $T = 5ma + 5mg = \frac{5mg}{11} + 5mg = \frac{60mg}{11}$ [2]

c) Force on pulley $= 2T = \frac{120mg}{11}$ [2]

d) $s = \frac{1}{2}at^2 = \frac{1}{2} \times \frac{g}{11} \times 2^2 = \frac{2g}{11} = 1.78\,\text{m}$ [2]

8. a) Newton 2   A: $T = 5ma$ [2]

Newton 2   B: $6mg - T = 6ma$ [2]

Adding equations: $6mg = 11ma$ [1]

$\Rightarrow a = \frac{6g}{11}$ [1]

b) $T = 5ma = 5m\left(\frac{6g}{11}\right) = \frac{30mg}{11}$ [2]

c) Force on pulley $= T\sqrt{2} = \frac{30\sqrt{2}mg}{11}$ [2]

d) A travels 1.5 m:

$s = \frac{1}{2}at^2 \Rightarrow t = \sqrt{\frac{2s}{a}} = \sqrt{\frac{33}{6g}} = 0.749$ seconds [2]

Answers

Types of Mark:

M = correct method applied to appropriate numbers
A = accuracy, dependent on **M** mark being gained
B = correct final answer, partially correct answer or correct intermediate stage

PRACTICE EXAM PAPER 1: PURE MATHEMATICS

1. $2x^2 - 3x + 5 = 8 - 2x$

$2x^2 - x - 3 = 0$ **M1**
$(2x - 3)(x + 1) = 0$ **M1**

$x = \dfrac{3}{2}$ or $x = -1$ **A1**

$x = \dfrac{3}{2}, y = 5$ or $x = -1, y = 10$ (correctly paired)   **A1**

2. $\dfrac{\sqrt{3}-1}{\sqrt{3}+1}\left(\dfrac{\sqrt{3}-1}{\sqrt{3}-1}\right) = \dfrac{3-2\sqrt{3}+1}{2} = \dfrac{4-2\sqrt{3}}{2} = 2-\sqrt{3}$

     M1 A1

$\dfrac{6}{\sqrt{3}} = \dfrac{6\sqrt{3}}{3} = 2\sqrt{3}$ **A1**

$2 - \sqrt{3} + 2\sqrt{3} = 2 + \sqrt{3}$ **A1**

3. $f(x) = \int f'(x) = \int \left(x^3 - 3x^2 + x + 6\right)\,dx$

$= \dfrac{x^4}{4} - x^3 + \dfrac{x^2}{2} + 6x + c$ **M1 A1**

Substituting in the point $(-2, -5)$:   **M1**
$4 + 8 + 2 - 4 + c = -5$
$2 + c = -5$
$c = -7$ **A1**

$\left(y = \dfrac{x^4}{4} - x^3 + \dfrac{x^2}{2} + 6x + -7\right)$ **A1**

4. (a) Substituting $x = 4$:   **M1**
$y = 3 - 2 = 1$ **A1**

(b) $y = 3 - 4x^{-\frac{1}{2}}$

$\dfrac{dy}{dx} = 2x^{-\frac{3}{2}}$ **M1 A1**

$x = 4 \Rightarrow \dfrac{dy}{dx} = \dfrac{1}{4}$ **A1**

$m = -4$ **M1**

$\dfrac{y-1}{x-4} = -4$ **M1**

$y - 1 = -4x + 16$
$4x + y - 17 = 0$ **A1**

5. $(3 - 2x)^9$

$= 3^9 + \dbinom{9}{1}3^8(-2x) + \dbinom{9}{2}3^7(-2x)^2 + \dbinom{9}{3}3^6(-2x)^3 + \dots$ **M1 A1**

$= 19\,683 - 118\,098x + 314\,928x^2$ **A4**

6. No real roots $\Rightarrow b^2 - 4ac < 0$ **M1**

$(k + 1)^2 - 12 < 0$ **A1**

$(k + 1)^2 - \left(2\sqrt{3}\right)^2 < 0$ **M1**

$\left((k+1) + 2\sqrt{3}\right)\left((k+1) - 2\sqrt{3}\right) < 0$ **M1**

Or: $k^2 + 2k - 11 < 0$ and quadratic formula   **M2**

Then:
Critical values are $k = -1 - 2\sqrt{3}$ and $k = 2\sqrt{3} - 1$ **A1**

$\therefore -1 - 2\sqrt{3} < k < -1 + 2\sqrt{3}$

7. $\log_6 x + \log_6(x - 9) = 2$

$\log_6 x(x - 9) = 2$ **M1**

$x(x - 9) = 6^2 = 36$ **A1**

$x^2 - 9x = 36$
$x^2 - 9x - 36 = 0$
$(x + 3)(x - 12) = 0$ **M1**

$x = -3$ or $x = 12$ **A2**

*Reject $x = -3$ solution since cannot take log of a negative number.   **B1**

8. $\dfrac{\sin C}{13} = \dfrac{\sin 135}{25}$ **M1**

$C = 21.57°$ **A1**

$\angle A = 180 - 135 - 21.57 = 23.43°$ **A1**

Area $= \dfrac{1}{2} \times 13 \times 25 \times \sin 23.43$ **M1**

$= 64.6\text{cm}^2$ **A1**

9. (a) For example when $A = B = 30°$, the left-hand side of the identity $= \tan 60 = \sqrt{3}$, but the right-hand side is $2\tan 30 = \dfrac{2\sqrt{3}}{3}$ **M1 A1**

(b) True when $A = 30°, B = 0°$ for example.   **M1 A1**

10. $10\cos^2 x + \sin x - 8 = 0$

$10\left(1 - \sin^2 x\right) + \sin x - 8 = 0$ **M1**

$10 - 10\sin^2 x + \sin x - 8 = 0$

$10\sin^2 x - \sin x - 2 = 0$ **A1**

Attempt to factorise or use quadratic formula: **M1**
$(2\sin x - 1)(5\sin x + 2) = 0$ **A1**

$\sin x = \dfrac{1}{2}$ or $\sin x = -\dfrac{2}{5}$ **A1**

$\sin x = \dfrac{1}{2} \Rightarrow x = 30°, 150°$ **A1**

$\sin x = -\dfrac{2}{5} \Rightarrow x = -23.6°, x = -156.4°, x = 203.6°,$

$x = 336.4°$ **A2**

11. $AB = OB - AB = 7\mathbf{i} + 10\mathbf{j} - (3\mathbf{i} + 12\mathbf{j}) = 4\mathbf{i} - 2\mathbf{j}$ **M1 A1**

Unit vector in this direction is $\dfrac{4\mathbf{i} - 2\mathbf{j}}{\sqrt{20}}$ **M1**

Vector of magnitude 10 in this direction is $\dfrac{10(4\mathbf{i} - 2\mathbf{j})}{\sqrt{20}}$

M1

$= \sqrt{5}(4\mathbf{i} - 2\mathbf{j})$ **A1**

Or:

Vector $4k\mathbf{i} - 2k\mathbf{j}$ has magnitude 10 **M1**

so $(4k)^2 + (2k)^2 = 100\left(\Rightarrow k = \sqrt{5}\right)$ **M1**

to give the vector $4\sqrt{5}\mathbf{i} - 2\sqrt{5}\mathbf{j}$ **A1**

12. Consider $\dfrac{dy}{dx} \approx \dfrac{2(x + h)^3 - 2x^3}{h}$ **M1**

$= \dfrac{2\left(x^3 + 3x^2h + 3xh^2 + h^3\right) - 2x^3}{h}$ **A1**

$= \dfrac{6x^2h + 6xh^2 + 2h^3}{h}$ **A1**

$= 6x^2 + 6xh + 2h^2$ **A1**

As $h \to 0, \dfrac{dy}{dx} \to 6x^2$ **A1**

13. $\dfrac{19 - 5x}{x - 3} = \dfrac{4 - 5(x - 3)}{x - 3} = \dfrac{4}{x - 3} - 5$ **M1 A1**

So asymptotes at $x = 3$ and $y = -5$. **A2**

Crosses axes at $\left(0, -\dfrac{19}{3}\right)$ and $\left(\dfrac{19}{5}, 0\right)$ **A2**

A1

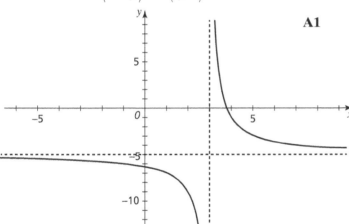

14. (a) (i) $f(-4) = 64 - 16 - 40 - 8 = 0$ **M1**

∴ $(x + 4)$ is a factor **A1**

(ii) $f(x) = (x + 4)\left(-x^2 + kx - 2\right)$

Equating coefficients of x^2 gives

$-4 + k = -1 \Rightarrow k = 3$ **M1 A1**

∴ $f(x) = (x + 4)\left(-x^2 + 3x - 2\right)$

∴ $f(x) = (x + 4)(2 - x)(x - 1)$ **M1 A1**

(b) **B1**

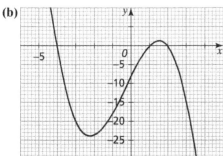

(c)

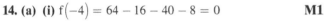

M1 A1 A1

15. (a) $4x^2 + 4y^2 - 12x - 24y + 41 = 0$

$x^2 + y^2 - 3x - 6y + \dfrac{41}{4} = 0$ **M1**

$x^2 - 3x + y^2 - 6y + \dfrac{41}{4} = 0$

$\left(x - \dfrac{3}{2}\right)^2 - \dfrac{9}{4} + (y - 3)^2 - 9 + \dfrac{41}{4} = 0$ **M1**

$\left(x - \dfrac{3}{2}\right)^2 + (y - 3)^2 = 1$ **A1**

Centre of circle has coordinates $\left(\dfrac{3}{2}, 3\right)$ and

radius $= 1$ **A2**

(b) Each tangent line is of the form $y = kx$

Substituting into circle equation:

$4x^2 + 4k^2x^2 - 12x - 24kx + 41 = 0$ **M1**

$\left(4k^2 + 4\right)x^2 - (24k + 12)x + 41 = 0$

Tangent lines, so each touch circle at one point

$\Rightarrow b^2 - 4ac = 0$ **M1**

$(12(2k + 1))^2 - 656\left(k^2 + 1\right) = 0$ **A1**

$9(2k + 1)^2 - 41\left(k^2 + 1\right) = 0$

$36k^2 + 36k + 9 - 41k^2 - 41 = 0$

$5k^2 - 36k + 32 = 0$

Using quadratic formula: **M1**

$k = 6.16$ or $= 1.04$ **A1**

Tangent lines are $y = 6.16x$ and $y = 1.04x$

16. (a) Reflection in the y-axis. **B1**

Stretch, scale factor $\dfrac{1}{2}$, parallel to x-axis. **B1**

Translation of 1 unit in direction of y-axis. **B1**

(b) Area $\displaystyle\int_{-\ln 3}^{\ln 2} \left(e^{-2x} + 1\right) dx$ **M1**

$= \left[-\dfrac{1}{2}e^{-2x} + x\right]_{-\ln 3}^{\ln 2}$ **A1 A1**

$= -\dfrac{1}{2}e^{-2\ln 2} + \ln 2 - \left(-\dfrac{1}{2}e^{2\ln 3} - \ln 3\right)$ **M1**

$= -\dfrac{1}{8} + \dfrac{9}{2} + \ln 6$

$= \dfrac{35}{8} + \ln 6$ **A1**

$\left(a = \dfrac{35}{8}, b = 1\right)$

Answers

PRACTICE EXAM PAPER 2: STATISTICS AND MECHANICS

Section A: Statistics

1. Considering the mean:

$$\frac{a + b + 12}{6} = \frac{8}{3} \Rightarrow a + b = 4 \qquad \textbf{M1 A1}$$

Considering the variance:

$$\frac{42 + a^2 + b^2}{6} - \left(\frac{8}{3}\right)^2 = \frac{14}{9} \qquad \textbf{M1 A1}$$

$$\Rightarrow a^2 + b^2 = 10$$

Solving simultaneously: **M1**
$$\Rightarrow a^2 + (4 - a)^2 = 10$$

$$\Rightarrow 2a^2 - 8a + 16 = 10$$

$$\Rightarrow a^2 - 4a + 3 = 0$$

$$\Rightarrow (a - 3)(a - 1) = 0$$

So $a = 3$ and $b = 1$, or $a = 1$ and $b = 3$ **A1**

Since $a < b$, the latter is true.

2. **(a)** If a student takes physics, then they cannot be taking economics, and vice versa. Therefore these events are dependent, hence not independent. **B1**

   (b) If independent, then $P(M \cap P) = P(M)P(P)$ **M1**

   So $\frac{1}{10} = \frac{6}{10}\left(p + \frac{1}{10}\right)$

   Solve to give $p = \frac{1}{15}$ **A1**

   (c) Summing all probabilities:
   $$\frac{1}{15} + \frac{1}{10} + \frac{4}{10} + \frac{1}{10} + \frac{2}{15} + q = 1 \qquad \textbf{M1}$$

   So $q = \frac{1}{5}$ **A1**

3. Re-ordering numbers:
   $9, 12, 13, 15, 16, 17, 18, 18, 19, 27, 31$
   $\sum f = 11$

   $Q_1 = 13, Q_2 = 17, Q_3 = 19$ **B1 B1 B1**

   $\frac{3}{2}(Q_3 - Q_1) = 9$

   $19 + 9 = 28$

   So 31 is the only outlier. **B1**
   Box plot: **B1**

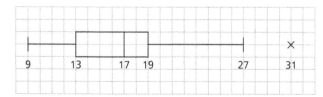

4. **(a)** From the diagram, 12.5 °C and 17.5 °C relate to 20 days and 28 days respectively. **M1**
   The estimate is therefore $28 - 20 = 8$ days **A1**

   (b) From the diagram, 10.5 °C relates to 8 days **M1**
   The probability is therefore $\frac{8}{30}$ or 0.27 **A1**

   (c) The lowest temperature that occurred cannot be obtained from the given diagram. **B1**

5. **(a) (i)** $X \sim B(0.12, 20)$ **M1**

   $$P(X = 3) = \binom{20}{3}(0.12^3)(0.88^{17}) = 0.224 \qquad \textbf{A1}$$

   (ii) $P(X \geq 3) = 1 - P(X \leq 2)$
   $$= 1 - P(X = 0) - P(X = 1) - P(X = 2) \qquad \textbf{M1}$$

   $$= 1 - 0.88^{20} - \binom{20}{1}(0.12)(0.88)^{19} - \binom{20}{2}(0.12)^2(0.88)^{18}$$

   $$= 0.437 \qquad \textbf{A1}$$

   (b) $H_0 : p = 0.12$ **B1**
   $H_1 : p > 0.12$ **B1**

   $P(X \geq 5) = 1 - P(X \leq 4) = 0.08$ **M1 A1**

   > 0.05

   So there is insufficient evidence to reject H_0 **A1**

Section B: Mechanics

6. **(a)** Velocity (ms^{-1}) **B1**

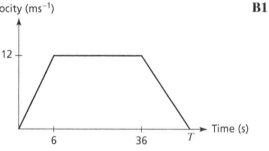

   (b) $T - 36 = \frac{12}{0.8} \Rightarrow T = 51$ seconds **M1 A1**

   (c) Using distance = area under graph:
   $$\frac{1}{2} \times (51 + 30) \times 12 = 486 \text{ metres} \qquad \textbf{M1 A1}$$

7. **(a)** Using $v^2 = u^2 + 2as$, $70^2 = 0^2 + 2 \times 2.5 \times s$ **M1**
   $s = 980$ metres **A1**

   (b) Distance travelled = 490 m

   Using $s = ut + \frac{1}{2}at^2$ **M1**

   $490 = \frac{1}{2} \times 2.5 \times t^2$

   $t = 19.8$ seconds **A1**

8. **(a)** $v = \int a\,dt = \int (4 - 2t)dt = 4t - t^2 + c$ **M1**

   At $t = 0, v = \frac{3}{2} \Rightarrow c = \frac{3}{2}$

   $v = \frac{3}{2} + 4t - t^2$ **A1**

   (b) At $t = 5, v = -\frac{7}{2}$ **M1**

   $\therefore$ speed $= \frac{7}{2}$ ms^{-1} **A1**

(c) Attempt to solve $\frac{3}{2} + 4t - t^2 = 0$ using
    quadratic formula (or equivalent)   **M1**
    $\Rightarrow t = 4.345$ seconds   **A1**

(d) $s = \displaystyle\int_0^{4.345} \frac{3}{2} + 4t - t^2 \, dt - \int_{4.345}^5 \frac{3}{2} + 4t - t^2 \, dt$ **M1**

$= \left[\frac{3t}{2} + 2t^2 - \frac{t^3}{3} \right]_0^{4.345} - \left[\frac{3t}{2} + 2t^2 - \frac{t^3}{3} \right]_{4.345}^5$ **A1**

$= 16.93 - \left(15.833 - 16.93 \right)$

$= 18.0$ m   **A1**

9. (a) A: $5mg - T = 5ma$ **M1 A1**
    B: $T - 2mg = 2ma$ **A1**

(b) Solving simultaneously: **M1**
    $3mg = 7ma$

    $a = \frac{3g}{7}$ **A1**

(c) $T = 2ma + 2mg = \frac{6mg}{7} + 2mg = \frac{20mg}{7}$ **A1**

(d) When A hits ground (using $v^2 = u^2 + 2as$)

    $v^2 = 0 + 2 \times \frac{3g}{7} \times 3$ **M1**

    $v = \sqrt{\frac{18g}{7}} \left(= 3\sqrt{\frac{2g}{7}} \right)$ **A1**

    Applying $v^2 = u^2 + 2as$ to B: **M1**

    $0 = \frac{18g}{7} - 2gs$

    $s = \frac{9}{7}$ **A1**

    Maximum height $= 3 + \frac{9}{7} = \frac{30}{7}$ m   **A1**

(e) Assumed tension is same either side of the
    pulley   **B1**

Acknowledgements

The authors and publisher are grateful to the copyright holders for permission to use quoted materials and images.

Pages 93, 100, 124, 126, 184, 213: Contains public sector information licensed under the Open Government Licence v1.0 All other images © Shutterstock.com

Every effort has been made to trace copyright holders and obtain their permission for the use of copyright material. The author and publisher will gladly receive information enabling them to rectify any error or omission in subsequent editions. All facts are correct at time of going to press.

Published by Collins
An imprint of HarperCollins*Publishers* Ltd
1 London Bridge Street
London SE1 9GF

HarperCollins*Publishers*
Macken House, 39/40 Mayor Street Upper,
Dublin 1, D01 C9W8, Ireland

© HarperCollins*Publishers* Limited 2020

ISBN 9780008268510

First published 2018
This edition published 2020

10 9 8 7 6 5

All rights reserved. No part of this publication may be reproduced, stored in a retrieval system, or transmitted, in any form or by any means, electronic, mechanical, photocopying, recording or otherwise, without the prior permission of Collins.

British Library Cataloguing in Publication Data.

A CIP record of this book is available from the British Library.

Authors: Rebecca Evans, Phil Duxbury and Leisa Bovey
Project management: Richard Toms
Commissioning: Katherine Wilkinson, Clare Souza and Kerry Ferguson
Cover Design: Sarah Duxbury and Kevin Robbins
Inside Concept Design: Ian Wrigley
Text Design and Layout: Jouve India Private Limited
Production: Natalia Rebow
Printed by Ashford Colour Press Ltd

MIX
Paper | Supporting responsible forestry
FSC
www.fsc.org
FSC™ C007454

This book contains FSC™ certified paper and other controlled sources to ensure responsible forest management.

For more information visit: www.harpercollins.co.uk/green